Recent Advances on Quasi-Metric Spaces

Recent Advances on Quasi-Metric Spaces

Special Issue Editors
Andreea Fulga
Erdal Karapinar

MDPI • Basel • Beijing • Wuhan • Barcelona • Belgrade

Special Issue Editors

Andreea Fulga
Department of Mathematics
and Computer Sciences,
Universitatea
Transilvania Brasov
Romania

Erdal Karapinar
Department of Medical
Research, China Medical
University Hospital,
China Medical University
Taiwan

Editorial Office
MDPI
St. Alban-Anlage 66
4052 Basel, Switzerland

This is a reprint of articles from the Special Issue published online in the open access journal *Mathematics* (ISSN 2227-7390) from 2019 to 2020 (available at: https://www.mdpi.com/journal/mathematics/special_issues/Recent_Advances_Quasi_Metric_Spaces).

For citation purposes, cite each article independently as indicated on the article page online and as indicated below:

LastName, A.A.; LastName, B.B.; LastName, C.C. Article Title. *Journal Name* **Year**, *Article Number*, Page Range.

ISBN 978-3-03928-881-6 (Pbk)
ISBN 978-3-03928-882-3 (PDF)

© 2020 by the authors. Articles in this book are Open Access and distributed under the Creative Commons Attribution (CC BY) license, which allows users to download, copy and build upon published articles, as long as the author and publisher are properly credited, which ensures maximum dissemination and a wider impact of our publications.

The book as a whole is distributed by MDPI under the terms and conditions of the Creative Commons license CC BY-NC-ND.

Contents

About the Special Issue Editors vii

Preface to "Recent Advances on Quasi-Metric Spaces" ix

Erdal Karapınar, Farshid Khojasteh and Zoran D. Mitrović
A Proposal for Revisiting Banach and Caristi Type Theorems in b-Metric Spaces
Reprinted from: *Mathematics* **2019**, *7*, 308, doi:10.3390/math7040308 1

Pradip R. Patle, Deepesh Kumar Patel, Hassen Aydi, Dhananjay Gopal and Nabil Mlaiki
Nadler and Kannan Type Set Valued Mappings in M-Metric Spaces and an Application
Reprinted from: *Mathematics* **2019**, *7*, 373, doi:10.3390/math7040373 5

Wasfi Shatanawi and Kamaleldin Abodayeh
Common Fixed Point under Nonlinear Contractions on Quasi Metric Spaces
Reprinted from: *Mathematics* **2019**, *7*, 453, doi:10.3390/math7050453 19

Obaid Alqahtani, Venigalla Madhulatha Hima Bindu and Erdal Karapınar
On Pata–Suzuki-Type Contractions
Reprinted from: *Mathematics* **2019**, *7*, 720, doi:10.3390/math7080720 28

Ekber Girgin, Mahpeyker Öztürk
Modified Suzuki-Simulation Type Contractive Mapping in Non-Archimedean Quasi Modular Metric Spaces and Application to Graph Theory
Reprinted from: *Mathematics* **2019**, *7*, 769, doi:10.3390/math7090769 39

Antonio Francisco Roldán López de Hierro and Naseer Shahzad
Ample Spectrum Contractions and Related Fixed Point Theorems
Reprinted from: *Mathematics* **2019**, *7*, 1033, doi:10.3390/math7111033 53

Salvador Romaguera and Pedro Tirado
A Characterization of Quasi-Metric Completeness in Terms of α–ψ-Contractive Mappings Having Fixed Points
Reprinted from: *Mathematics* **2020**, *8*, 16, doi:10.3390/math8010016 76

Watcharin Chartbupapan, Ovidiu Bagdasar and Kanit Mukdasai
A Novel Delay-Dependent Asymptotic Stability Conditions for Differential and Riemann-Liouville Fractional DifferentialNeutral Systems with Constant Delays and Nonlinear Perturbation
Reprinted from: *Mathematics* **2020**, *8*, 82, doi:10.3390/math8010082 81

About the Special Issue Editors

Andreea Fulgais a lecturer at the Department of Mathematics and Computer Science at Transylvania University in Brasov, Romania. She received her Ph.D. in mathematics (2007) from Transylvania University of Brasov, with the thesis "Limit Theorems, Convergence Problems and Applications". Her research interests include functional analysis, operator theory, and fixed-point theory. She is the author of five books and over 50 publications in journals, book chapters, or conference proceedings and acts as a reviewer for various journals. She is a member of the Mathematical Sciences Society of Romania, has won certain national competitions, and has participated as a member in some research projects.

Erdal Karapinar is currently a visiting professor at China Medical University, Taichung, Taiwan. He received his Ph.D. in mathematics, in particular, functional analysis, at Middle East Technical University in 2014 under the supervision of Professor V. P. Zakharyuta (Sabancı University) and Prof. M. H. Yurdakul (Middle East Technical University). He was a post-doc researcher at Sabanci University from 2014 to 2015. After his military service, he worked in Izmir University of Economics, from 2015 to 2017. Later, he worked at Atilim University from 2017 to 2019. He has been a full professor in mathematics since December 2011. From January 2019 to the present, he has been a visiting professor at China Medical University. He has published more than 350 papers and a monograph. He is currently acting as an associate editor of more than 10 journals. He is also a founder and Editor-in-Chief of two journals, *Advances in the Theory of Nonlinear Analysis and its Applications* (*ATNAA*) and *Results in Nonlinear Analysis* (*RNA*). He is a highly cited researcher in mathematics, according to the Clarivate Analysis, from 2014 to 2019. He was also selected as a top reviewer in mathematics by *Publons* (which belongs to Clarivate Analysis).

Preface to "Recent Advances on Quasi-Metric Spaces"

If we were to say that fixed-point theory appeared in Liouville's article on solutions of differential equations (1837) in the second quarter of the 18th century, it would not be wrong. This approach was further developed by Picard in 1890 and entered the literature as a method of successive approximations. This method was abstracted and extracted as a separate fixed-point theorem in the setting of complete normed space by Banach in 1922.

For this reason, usually, it is said that fixed-point theory was founded by Banach. In its earlier iteration, this first fixed-point theorem was known as the Picard–Banach theorem. Later, the analog of that theorem was proved in the framework of complete metric spaces by Caccioppoli in 1931. In some literature, the Banach–Caccioppoli theorem is indicated as a first fixed-point theorem in the setting of a complete metric space.

As we mentioned above, fixed-point theory can be considered as a theory that was derived from applied mathematics. On the other hand, the techniques belong to functional analysis and topology.

In particular, this theory, and its potential application, has been investigated and focused on by a great number of researchers. It should be underlined that this theory has been applied in physics, economics, engineering, computer science, and so on. Indeed, an application for fixed-point theorem can be found in all fields of quantitative science.

In this Special Issue, we focused on fixed-point results in the setting of quasi-metric spaces and applications but were not restricted to it. The selected papers express our aims in this regard.

Andreea Fulga, Erdal Karapinar
Special Issue Editors

Article

A Proposal for Revisiting Banach and Caristi Type Theorems in *b*-Metric Spaces

Erdal Karapınar [1,*], **Farshid Khojasteh** [2] **and Zoran D. Mitrović** [3]

1. Department of Medical Research, China Medical University Hospital, China Medical University, Taichung 40402, Taiwan
2. Young Researcher and Elite Club, Arak Branch, Islamic Azad University, Arak 38361-1-9131, Iran; fr_khojasteh@yahoo.com
2. Faculty of Electrical Engineering, University of Banja Luka, 78000 Banja luka, Bosnia and Herzegovina; zoran.mitrovic@etf.unibl.org
* Correspondence: karapinar@mail.cmuh.org.tw or erdalkarapinar@yahoo.com

Received: 26 February 2019; Accepted: 20 March 2019; Published: 27 March 2019

Abstract: In this paper, we revisit the renowned fixed point theorems belongs to Caristi and Banach. We propose a new fixed point theorem which is inspired from both Caristi and Banach. We also consider an example to illustrate our result.

Keywords: *b*-metric; Banach fixed point theorem; Caristi fixed point theorem

MSC: 46T99; 47H10; 54H25

---

1. Introduction and Preliminaries

In fixed point theory, the approaches of the renowned results of Caristi [1] and Banach [2] are quite different and the structures of the corresponding proofs varies. In this short note, we propose a new fixed point theorem that is inspired from these two famous results.

We aim to present our results in the largest framework, *b*-metric space, instead of standard metric space. The concept of *b*-metric has been discovered several times by different authors with distinct names, such as quasi-metric, generalized metric and so on. On the other hand, this concept became popular after the interesting papers of Bakhtin [3] and Czerwik [4]. For more details in b-metric space and advances in fixed point theory in the setting of *b*-metric spaces, we refer e.g., [5–17].

Definition 1. *Let X be a nonempty set and $s \geq 1$ be a real number. We say that $d : X \times X \to [0,1)$ is a b-metric with coefficient s when, for each $x, y, z \in X$,*

(b1) $d(x,y) = d(y,x)$;
(b2) $d(x,y) = 0$ if and only if $x = y$;
(b3) $d(x,z) \leq s[d(x,y) + d(y,z)]$ *(Expanded triangle inequality).*

In this case, the triple (X, d, s) is called a b-metric space with coefficient s.

The classical examples and crucial examples of *b*-metric spaces are $l^p(\mathbb{R})$ and $L^p[0,1]$, $p \in (0,1)$.

The topological notions (such as, convergence, Cauchy criteria, completeness, and so on) are defined by verbatim of the corresponding notions for standard metric. On the other hand, we should underline the fact that *b*-metric does need to be continuous, for certain details, see e.g., [3,4].

We recollect the following basic observations here.

Lemma 1. *[14] For a sequence $(\theta_n)_{n\in\mathbb{N}}$ in a b-metric space (X,d,s), there exists a constant $\gamma \in [0,1)$ such that*

$$d(\theta_{n+1},\theta_n) \leq \gamma d(\theta_n,\theta_{n-1}), \text{ for all } n \in \mathbb{N}.$$

Then, the sequence $(\theta_n)_{n\in\mathbb{N}}$ is fundamental (Cauchy).

The aim of this paper is to correlate the Banach type fixed point result with Caristi type fixed point results in b-metric spaces.

2. Main Result

Theorem 1. *Let (X,d,s) be a complete metric space and $T: X \to X$ be a map. Suppose that there exists a function $\varphi: X \to \mathbb{R}$ with*

(i) *φ is bounded from below ($\inf \varphi(X) > -\infty$),*
(ii) *$d(x,Tx) > 0$ implies $d(Tx,Ty) \leq (\varphi(x) - \phi(Tx))d(x,y)$.*

Then, T has at least one fixed point in X.

Proof. Let $\theta_0 \in X$. If $T\theta_0 = \theta_0$, the proof is completed. Herewith, we assume $d(\theta_0, T\theta_0) > 0$. Without loss of generality, keeping the same argument in mind, we assume that $\theta_{n+1} = T\theta_n$ and hence

$$d(\theta_n, \theta_{n+1}) = d(\theta_n, T\theta_n) > 0. \qquad (1)$$

For that sake of convenience, suppose that $a_n = d(\theta_n, \theta_{n-1})$. From (ii), we derive that

$$\begin{aligned} a_{n+1} &= d(\theta_n, \theta_{n+1}) = d(T\theta_{n-1}, T\theta_n) \\ &\leq (\varphi(\theta_{n-1}) - \varphi(T\theta_{n-1}))d(\theta_{n-1}, \theta_n) \\ &= (\varphi(\theta_{n-1}) - \varphi(\theta_n))a_n. \end{aligned}$$

So we have,

$$0 < \frac{a_{n+1}}{a_n} \leq \varphi(\theta_{n-1}) - \varphi(\theta_n) \text{ for each } n \in \mathbb{N}.$$

Thus the sequence $\{\varphi(\theta_n)\}$ is necessarily positive and non-increasing. Hence, it converges to some $r \geq 0$. On the other hand, for each $n \in \mathbb{N}$, we have

$$\begin{aligned} \sum_{k=1}^n \frac{a_{k+1}}{a_k} &\leq \sum_{k=1}^n (\varphi(\theta_{k-1}) - \varphi(\theta_k)) \\ &= (\varphi(\theta_0) - \varphi(\theta_1)) + (\varphi(\theta_1) - \varphi(\theta_2)) + \ldots + (\varphi(\theta_{n-1}) - \varphi(\theta_n)) \\ &= \varphi(\theta_0) - \varphi(\theta_n) \to \varphi(\theta_0) - r < \infty, \text{ as } n \to \infty. \end{aligned}$$

It means that

$$\sum_{n=1}^\infty \frac{a_{n+1}}{a_n} < \infty.$$

Accordingly, we have

$$\lim_{n\to\infty} \frac{a_{n+1}}{a_n} = 0. \qquad (2)$$

On account of (2), for $\gamma \in (0,1)$, there exists $n_0 \in \mathbb{N}$ such that

$$\frac{a_{n+1}}{a_n} \leq \gamma, \qquad (3)$$

for all $n \geq n_0$. It yields that

$$d(\theta_{n+1},\theta_n) \leq \gamma d(\theta_n, \theta_{n-1}), \qquad (4)$$

for all $n \geq n_0$. Now using Lemma 1 we obtain that the sequence $\{\theta_n\}$ converges to some $\omega \in X$. We claim that ω is the fixed point of T. Employing assumption (ii) of the theorem, we find that

$$\begin{aligned} d(\omega, T\omega) &\leq s[d(\omega, \theta_{n+1}) + d(\theta_{n+1}, T\omega)] \\ &\leq s[d(\omega, \theta_{n+1}) + (\varphi(\theta_n) - \varphi(\omega))d(\theta_n, \omega)] \to 0 \text{ as } n \to \infty. \end{aligned}$$

Consequently, we obtain $d(\omega, T\omega) = 0$, that is, $T\omega = \omega$. □

From Theorem 1, we get the corresponding result for complete metric spaces. The following example shows that the Theorem 1 is not a consequence of Banach's contraction principle.

Example 1. Let $X = \{0, 1, 2\}$ endowed with the following metric:

$$d(0,1) = 1, d(2,0) = 1, d(1,2) = \frac{3}{2} \text{ and } d(a,a) = 0, \text{ for all } a \in X, \ d(a,b) = d(b,a), \text{ for all } a, b \in X.$$

Let $T(0) = 0, T(1) = 2, T(2) = 0$. Define $\varphi : X \to [0, \infty)$ as $\varphi(2) = 2, \varphi(0) = 0, \varphi(1) = 4$. Thus for all $x \in X$ such that $d(x, Tx) > 0$, (in this example, $x \neq 0$), we have

$$\begin{aligned} d(T1, T2) &\leq (\varphi(1) - \varphi(T(1)))d(2,1), \\ d(T2, T1) &\leq (\varphi(2) - \varphi(T(2)))d(2,1), \\ d(T1, T0) &\leq (\varphi(1) - \varphi(T(1)))d(1,0), \\ d(T2, T0) &\leq (\varphi(2) - \varphi(T(2)))d(2,0). \end{aligned}$$

Thus the mapping T satisfies our condition and also has a fixed point. Note that $d(T1, T0) = d(1, 0)$. Thus, it does not satisfy the Banach contraction principle.

Remark 1.
1. From Example 1, it follows that Theorem 1 (over metric spaces) is not a consequence of the Banach contraction principle.
2. Question for further study: It is natural to ask if the Banach contraction principle is a consequence of Theorem 1 (over metric spaces).

Author Contributions: All authors contributed equally and significantly in writing this article. All authors read and approved the final manuscript.

Funding: This research received no external funding.

Conflicts of Interest: The authors declare no conflict of interest.

References

1. Caristi, J. Fixed point theorems for mappings satisfying inwardness conditions. *Trans. Am. Math. Soc.* **1976**, *215*, 241–251. [CrossRef]
2. Banach, B. Sur les opérations dans les ensembles abstraits et leur application aux équations intégrales. *Fundam. Math.* **1992**, *3*, 133–181. [CrossRef]
3. Bakhtin, I.A. The contraction mapping principle in quasimetric spaces. *Funct. Anal. Ulianowsk Gos. Ped. Inst.* **1989**, *30*, 26–37.
4. Czerwik, S. Contraction mappings in b-metric spaces. *Acta Math. Inform. Univ. Ostrav.* **1993**, *1*, 5–11.
5. Afshari, H.; Aydi, H.; Karapinar, E. Existence of Fixed Points of Set-Valued Mappings in b-metric Spaces. *East Asian Math. J.* **2016**, *32*, 319–332. [CrossRef]
6. Aksoy, U.; Karapinar, E.; Erhan, I.M. Fixed points of generalized alpha-admissible contractions on b-metric spaces with an application to boundary value problems. *J. Nonlinear Convex Anal.* **2016**, *17*, 1095–1108.
7. Alsulami, H.; Gulyaz, S.; Karapinar, E.; Erhan, I. An Ulam stability result on quasi-b-metric-like spaces. *Open Math.* **2016**, *14*, 1087–1103. [CrossRef]

8. Aydi, H.; Bota, M.F.; Karapinar, E.; Mitrović, S. A fixed point theorem for set-valued quasi-contractions in b-metric spaces. *Fixed Point Theory Appl.* **2012**, *2012*, 88. [CrossRef]
9. Aydi, H.; Bota, M.-F.; Karapinar, E.; Moradi, S. A common fixed point for weak-ϕ-contractions on b-metric spaces. *Fixed Point Theory* **2012**, *13*, 337–346.
10. Bota, M.-F.; Karapinar, E.; Mlesnite, O. Ulam-Hyers stability results for fixed point problems via alpha-psi-contractive mapping in b-metric space. *Abstr. Appl. Anal.* **2013**, *2013*, 825293. [CrossRef]
11. Bota, M.-F.; Karapinar, E. A note on "Some results on multi-valued weakly Jungck mappings in b-metric space". *Cent. Eur. J. Math.* **2013**, *11*, 1711–1712. [CrossRef]
12. Bota, M.; Chifu, C.; Karapinar, E. Fixed point theorems for generalized $(\alpha - \psi)$-Ciric-type contractive multivalued operators in b-metric spaces. *J. Nonlinear Sci. Appl.* **2016**, *9*, 1165–1177. [CrossRef]
13. Hammache, K.; Karapınar, E.; Ould-Hammouda, A. On Admissible weak contractions in b-metric-like space. *J. Math. Anal.* **2017**, *8*, 167–180.
14. Mitrović, Z.D. A note on the result of Suzuki, Miculescu and Mihail. *J. Fixed Point Theory Appl.* **2019**. [CrossRef]
15. Mitrović, Z.D.; Radenović, S. The Banach and Reich contractions in $b_v(s)$-metric spaces. *J. Fixed Point Theory Appl.* **2017**, *19*, 3087–3095. [CrossRef]
16. Mitrović, Z.D.; Radenović, S. A common fixed point theorem of Jungck in rectangular b-metric spaces. *Acta Math. Hungar.* **2017**, *153*, 401–407. [CrossRef]
17. Mitrović, Z.D. A note on a Banach's fixed point theorem in b-rectangular metric space and b-metric space. *Math. Slovaca* **2018**, *68*, 1113–1116. [CrossRef]

© 2019 by the authors. Licensee MDPI, Basel, Switzerland. This article is an open access article distributed under the terms and conditions of the Creative Commons Attribution (CC BY) license (http://creativecommons.org/licenses/by/4.0/).

Article

Nadler and Kannan Type Set Valued Mappings in M-Metric Spaces and an Application

Pradip R. Patle [1], Deepesh Kumar Patel [1], Hassen Aydi [2,3,*], Dhananjay Gopal [4] and Nabil Mlaiki [5]

1. Department of Mathematics, Visvesvaraya National Institute of Technology, Nagpur 440010, India; pradip.patle12@gmail.com (P.R.P.); deepesh456@gmail.com (D.K.P.)
2. Université de Sousse, Institut Supérieur d'Informatique et des Techniques de Communication, H. Sousse 4000, Tunisia
3. China Medical University Hospital, China Medical University, Taichung 40402, Taiwan
4. Department of Applied Mathematics & Humanities, S.V. National Institute of Technology, Surat 395007, Gujarat, India; gopaldhananjay@yahoo.in
5. Department of Mathematics and General Sciences, Prince Sultan University, P. O. Box 66833, Riyadh 11586, Saudi Arabia; nmlaiki@psu.edu.sa or nmlaiki2012@gmail.com
* Correspondence: hassen.aydi@isima.rnu.tn

Received: 27 February 2019; Accepted: 18 April 2019; Published: 24 April 2019

Abstract: This article intends to initiate the study of Pompeiu–Hausdorff distance induced by an M-metric. The Nadler and Kannan type fixed point theorems for set-valued mappings are also established in the said spaces. Moreover, the discussion is supported with the aid of competent examples and a result on homotopy. This approach improves the current state of art in fixed point theory.

Keywords: homotopy; M-metric; M-Pompeiu–Hausdorff type metric; multivalued mapping; fixed point

MSC: Primary 47H10; Secondary 54H25, 05C40

1. Introduction

With the introduction of Banach's contraction principle (BCP), the fixed point theory advanced in various directions. Nadler [1] obtained the fundamental fixed point result for set-valued mappings using the notion of Pompeiu–Hausdorff metric which is an extension of the BCP. Later on, many fixed point theorists followed the findings of Nadler and contributed significantly to the development of theory (cf. S. Reich [2,3]).

On the other hand, in order to investigate the semantics of data flow networks; Matthews [4] coined the concept called as partial metric spaces which are used efficiently while building models in computation theory. On the inclusion of partial metric spaces into literature, many fixed point theorems were established in this setting, see [5–16]. Recently, Asadi et al. [17] brought the notion of an M-metric as a real generalization of a partial metric into the literature. They also obtained the M-metric version of the fixed point results of Banach and Kannan. Also, some fixed point theorems have been established in M-metric spaces endowed with a graph, see [18].

In this work, we introduce the M-Pompeiu–Hausdorff type metric. Furthermore, we extend the fixed point theorems of Nadler and Kannan to M-metric spaces for set-valued mappings. Finally, homotopy results for M-metric spaces are discussed.

2. Preliminaries

The symbols $\mathbb{N}$, $\mathbb{R}$ and $\mathbb{R}^+$ represent respectively set of all natural numbers, real numbers and nonnegative real numbers. Let us recall some of the concepts for simplicity in understanding.

Definition 1 ([4]). *Let X be a nonempty set. Then a partial metric is a function $p : X \times X \to \mathbb{R}^+$ satisfying following conditions:*

(p_1) $a = b \iff p(a,a) = p(a,b) = p(b,b)$;
(p_2) $p(a,a) \leq p(a,b)$;
(p_3) $p(a,b) = p(b,a)$;
(p_4) $p(a,b) \leq p(a,c) + p(c,b) - p(c,c)$.

for all $a, b, c \in X$. The pair (X, p) is called a partial metric space.

The concept of an M-metric [17] defined in following definition extends and generalize the notion of partial metric.

Definition 2 ([17]). *Let X be a non empty set. Then an M-metric is a function $m : X \times X \to \mathbb{R}^+$ satisfying the following conditions:*

(m_1) $m(a,a) = m(b,b) = m(a,b) \Leftrightarrow a = b$;
(m_2) $m_{ab} \leq m(a,b)$ where $m_{ab} := \min\{m(a,a), m(b,b)\}$;
(m_3) $m(a,b) = m(b,a)$;
(m_4) $(m(a,b) - m_{ab}) \leq (m(a,c) - m_{ac}) + (m(c,b) - m_{cb})$.

for all $a, b, c \in X$. The pair (X, m) is called an M-metric space.

Remark 1 ([17]). *Let us denote $M_{ab} := \max\{m(a,a), m(b,b)\}$, where m is an M-metric on X. Then for every $a, b \in X$, we have*

(1) $0 \leq M_{ab} + m_{ab} = m(a,a) + m(b,b)$,
(2) $0 \leq M_{ab} - m_{ab} = |m(a,a) - m(b,b)|$,
(3) $M_{ab} - m_{ab} \leq (M_{ac} - m_{ac}) + (M_{cb} - m_{cb})$.

Example 1 ([17]). *Let m be an M-metric on X. Then*

(1) $m^w(a,b) = m(a,b) - 2m_{ab} + M_{ab}$,
(2) $m^s(a,b) = \begin{cases} m(a,b) - m_{ab} & \text{if } a \neq b, \\ 0 & \text{if } a = b, \end{cases}$

are ordinary metrics on X.

Two new examples of M-metrics are as follows:

Example 2. *Let $X = [0, \infty)$. Then*

(a) $m_1(a,b) = |a - b| + \frac{a+b}{2}$,
(b) $m_2(a,b) = |a - b| + \frac{a+b}{3}$

are M-metrics on X.

Let $B_m(a, \eta) = \{b \in X : m(a,b) < m_{ab} + \eta\}$ be the open ball with center a and radius $\eta > 0$ in M-metric space (X, m). The collection $\{B_m(a, \eta) : a \in X, \eta > 0\}$, acts as a basis for the topology τ_m (say) on M-metric X.

Remark 2 ([17]). τ_m is T_0 but not Hausdorff.

Definition 3 ([17]). Let $\{a_k\}$ be a sequence in M-metric spaces (X, m).

(1) $\{a_k\}$ is called M-convergent to $a \in X$ if and only if
$$\lim_{k \to \infty} (m(a_k, a) - m_{a_k a}) = 0.$$

(2) If $\lim_{k,j \to \infty} (m(a_k, a_j) - m_{a_k a_j})$ and $\lim_{k,j \to \infty} (M_{a_k a_j} - m_{a_k a_j})$ exist and finite then the sequence $\{a_k\}$ is called M-Cauchy.

(3) If every M-Cauchy sequence $\{a_k\}$ is M-convergent, with respect to τ_m, to $a \in X$ such that $\lim_{k \to \infty} (m(a_k, a) - m_{a_k a}) = 0$ and $\lim_{k \to \infty} (M_{a_k a} - m_{a_k a}) = 0$ then (X, m) is called M-complete.

Lemma 1 ([17]). Let $\{a_k\}$ be a sequence in M-metric spaces (X, m). Then

(i) $\{a_k\}$ is M-Cauchy if and only if it is a Cauchy sequence in the metric space (X, m^w).
(ii) (X, m) is M-complete if and only if (X, m^w) is complete.

Example 3. Let X and $m_1, m_2 : X \times X \to [0, \infty)$ be as defined in Example 2 for all $a, b \in X$. Then (X, m_1) and (X, m_2) are M-complete. Indeed, $(X, m^w) = ([0, \infty), k|x - y|)$ is a complete metric space, where $k = \frac{5}{2}$ for m_1 and $k = 2$ for m_2.

Lemma 2 ([17]). Let $a_k \to a$ and $b_k \to b$ as $k \to \infty$ in (X, m). Then as $k \to \infty$, $(m(a_k, b_k) - m_{a_k b_k}) \to (m(a, b) - m_{ab})$.

Lemma 3 ([17]). Let $a_k \to a$ as $k \to \infty$ in (X, m). Then $(m(a_k, b) - m_{a_k b}) \to (m(a, b) - m_{ab})$, $k \to \infty$, for all $b \in X$.

Lemma 4 ([17]). Let $a_k \to a$ and $a_k \to b$ as $k \to \infty$ in (X, m). Then $m(a, b) = m_{ab}$. Further, if $m(a, a) = m(b, b)$, then $a = b$.

Lemma 5 ([17]). Let $\{a_k\}$ be a sequence in (X, m) such that for some $r \in [0, 1)$, $m(a_{k+1}, a_k) \le rm(a_k, a_{k-1})$, $k \in \mathbb{N}$ then

(a) $\lim_{k \to \infty} m(a_k, a_{k-1}) = 0$;
(b) $\lim_{k \to \infty} m(a_k, a_k) = 0$;
(c) $\lim_{k,j \to \infty} m_{a_k, a_j} = 0$;
(d) $\{a_k\}$ is M-Cauchy.

3. M-Pompeiu–Hausdorff Type Metric

The concept of a partial Hausdorff metric is defined in [19,20]. Following them we initiate the notion of an M-Pompeiu–Hausdorff type metric induced by an M-metric in this section. Let us begin with the following definition.

Definition 4. A subset A of an M-metric space (X, m) is called bounded if for all $a \in A$, there exist $b \in X$ and $K \ge 0$ such that $a \in B_m(b, K)$, that is, $m(a, b) < m_{ba} + K$.

Let $\mathcal{CB}^m(X)$ denotes the family of all nonempty, bounded, and closed subsets in (X, m). For $P, Q \in \mathcal{CB}^m(X)$, define
$$\mathcal{H}_m(P, Q) = \max\{\delta_m(P, Q), \delta_m(Q, P)\},$$
where $\delta_m(P, Q) = \sup\{m(a, Q) : a \in P\}$ and $m(a, Q) = \inf\{m(a, b) : b \in Q\}$.

Let $\overline{P}$ denote the closure of P with respect to M-metric m. Note that P is closed in (X,m) if and only if $\overline{P} = P$.

Lemma 6. *Let P be any nonempty set in an M-metric space (X,m), then $a \in \overline{P}$ if and only if $m(a,P) = \sup_{x \in P} m_{ax}$.*

Proof.

$$\begin{aligned}
a \in \overline{P} &\Leftrightarrow B_m(a,\eta) \cap P \neq \emptyset, \text{ for all } \eta > 0 \\
&\Leftrightarrow m(a,x) < m_{ax} + \eta, \text{ for some } x \in P \\
&\Leftrightarrow m(a,x) - m_{ax} < \eta \\
&\Leftrightarrow \inf\{m(a,x) - m_{ax} : x \in P\} = 0 \\
&\Leftrightarrow \inf\{m(a,x) : x \in P\} = \sup\{m_{ax} : x \in P\} \\
&\Leftrightarrow m(a,P) = \sup_{x \in P} m_{ax}.
\end{aligned}$$

□

Proposition 1. *Let $P, Q, R \in \mathcal{CB}^m(X)$, then we have*

(a) $\delta_m(P,P) = \sup_{a \in P}\{\sup_{b \in P} m_{ab}\}$;

(b) $(\delta_m(P,Q) - \sup_{a \in P}\sup_{b \in Q} m_{ab}) \leq (\delta_m(P,R) - \inf_{a \in P}\inf_{c \in R} m_{ac}) + (\delta_m(R,Q) - \inf_{c \in R}\inf_{b \in Q} m_{cb})$.

Proof.

(a) Since $P \in \mathcal{CB}^m(X)$, $P = \overline{P}$. Then from Lemma 6, $m(a,P) = \sup_{x \in P} m_{ax}$. Therefore, $\delta_m(P,P) = \sup_{a \in P}\{m(a,P)\} = \sup_{a \in P}\{\sup_{x \in P} m_{ax}\}$.

(b) For any $a \in P$, $b \in Q$ and $c \in R$, we have

$$m(a,b) - m_{ab} \leq m(a,c) - m_{ac} + m(c,b) - m_{cb}.$$

We rewrite it as

$$m(a,b) - m_{ab} + m_{ac} + m_{cb} \leq m(a,c) + m(c,b).$$

Since b is arbitrary element in Q, we have

$$m(a,Q) - \sup_{b \in Q} m_{ab} + m_{ac} + \inf_{b \in Q} m_{cb} \leq m(a,c) + m(c,Q).$$

Since $m(c,Q) \leq \delta_m(R,Q)$, we can write above inequality as

$$m(a,Q) - \sup_{b \in Q} m_{ab} + m_{ac} + \inf_{b \in Q} m_{cb} \leq m(a,c) + \delta_m(R,Q).$$

As c is arbitrary in R, we have

$$m(a,Q) - \sup_{b \in Q} m_{ab} + \inf_{c \in R} m_{ac} + \inf_{c \in R}\inf_{b \in Q} m_{cb} \leq m(a,R) + \delta_m(R,Q).$$

We rewrite the above inequality as

$$m(a,Q) + \inf_{c \in R}\inf_{b \in Q} m_{cb} \leq m(a,R) + \delta_m(R,Q) + \sup_{b \in Q} m_{ab} - \inf_{c \in R} m_{ac}.$$

Again, as a is arbitrary in P, we get

$$\delta_m(P,Q) + \inf_{c\in R}\inf_{b\in Q} m_{cb} \leq \delta_m(P,R) + \delta_m(R,Q) + \sup_{a\in P}\sup_{b\in Q} m_{ab} - \inf_{a\in P}\inf_{c\in R} m_{ac}.$$

□

Proposition 2. *For any $P, Q, R \in \mathcal{CB}^m(X)$ following are true*

(i) $\mathcal{H}_m(P,P) = \delta_m(P,P) = \sup_{a\in P}\{\sup_{b\in P} m_{ab}\}$;

(ii) $\mathcal{H}_m(P,Q) = \mathcal{H}_m(Q,P)$;

(iii) $\mathcal{H}_m(P,Q) - \sup_{a\in P}\sup_{b\in Q} m_{ab} \leq \mathcal{H}_m(P,R) + \mathcal{H}_m(Q,R) - \inf_{a\in P}\inf_{c\in R} m_{ac} - \inf_{c\in R}\inf_{b\in Q} m_{cb}$.

Proof.

(i) From (a) of Proposition 1, we write $\mathcal{H}_m(P,P) = \delta_m(P,P) = \sup_{a\in P}\{\sup_{b\in P} m_{ab}\}$.

(ii) It follows from (m$_2$) of Definition 2.

(iii) Using (b) of Proposition 1, we have

$$\mathcal{H}_m(P,Q) = \max\{\delta_m(P,Q), \delta_m(Q,P)\}$$

$$\leq \max\left\{[\delta_m(P,R) - \inf_{a\in P}\inf_{c\in R} m_{ac} + \delta_m(R,Q) - \inf_{c\in R}\inf_{b\in Q} m_{cb} + \sup_{a\in P}\sup_{b\in Q} m_{ab}],\right.$$

$$\left.[\delta_m(Q,R) - \inf_{a\in P}\inf_{c\in R} m_{ac} + \delta_m(R,P) - \inf_{c\in R}\inf_{b\in Q} m_{cb} + \sup_{a\in P}\sup_{b\in Q} m_{ab}]\right\}$$

$$\leq \max\{\delta_m(P,R), \delta_m(R,P)\} + \max\{\delta_m(Q,R), \delta_m(R,Q)\}$$

$$- \inf_{a\in P}\inf_{c\in R} m_{ac} - \inf_{c\in R}\inf_{b\in Q} m_{cb} + \sup_{a\in P}\sup_{b\in Q} m_{ab}$$

$$\leq \mathcal{H}_m(P,R) + \mathcal{H}_m(R,Q) - \inf_{a\in P}\inf_{c\in R} m_{ac} - \inf_{c\in R}\inf_{b\in Q} m_{cb} + \sup_{a\in P}\sup_{b\in Q} m_{ab}.$$

□

Remark 3. *In general, $\mathcal{H}_m(A,A) \neq 0$ for $A \in \mathcal{CB}^m(X)$. It can be verified through the following example.*

Example 4. *Let $X = [0,\infty)$ and $m(a,b) = \frac{a+b}{2}$, then clearly (X,m) is an M-metric space. In view of (a) of Proposition 1, we have*

$$\mathcal{H}_m([1,2],[1,2]) = \delta_m([1,2],[1,2]) = \sup_{p\in[1,2]}\sup_{q\in[1,2]} m_{pq} = \sup_{p\in[1,2]}\sup_{q\in[1,2]} \min\{p,q\} \neq 0.$$

In view of Proposition 2, we call $\mathcal{H}_m : \mathcal{CB}^m(X) \times \mathcal{CB}^m(X) \to [0,+\infty)$ an M-Pompeiu–Hausdorff type metric induced by m.

Lemma 7. *Let $P, Q \in \mathcal{CB}^m(X)$ and $q > 1$. Then for every $a \in P$, there is at least one $b \in Q$ such that $m(a,b) \leq q\mathcal{H}_m(P,Q)$.*

Proof. Assume that there exists an $a \in P$ such that $m(a,b) > q\mathcal{H}_m(P,Q)$ for all $b \in Q$. This implies that

$$\inf_{b\in Q}\{m(a,b)\} \geq q\mathcal{H}_m(P,Q),$$

that is,

$$m(a,Q) \geq q\mathcal{H}_m(P,Q).$$

Note that
$$\mathcal{H}_m(P,Q) \geq \delta_m(P,Q) = \sup_{x \in P} m(x,Q) \geq m(a,Q) \geq q\mathcal{H}_m(P,Q).$$

Since $\mathcal{H}_m(P,Q) \neq 0$, $q \leq 1$, which is a contradiction. □

Lemma 8. *Let $P,Q \in \mathcal{CB}^m(X)$ and $r > 0$. For any $a \in P$, there is at least one $b \in Q$ such that $m(a,b) \leq \mathcal{H}_m(P,Q) + r$.*

Proof. Assume that there exists $a \in P$ such that $m(a,b) > \mathcal{H}_m(P,Q) + r$ for all $b \in Q$. This implies that
$$\inf_{b \in Q}\{m(a,b)\} \geq \mathcal{H}_m(P,Q) + r,$$

that is,
$$m(a,Q) \geq \mathcal{H}_m(P,Q) + r.$$

Now,
$$\mathcal{H}_m(P,Q) + r \leq m(a,Q) \leq \delta_m(P,Q) \leq \mathcal{H}_m(P,Q).$$

Thus, $r \leq 0$, which is a contradiction. □

4. Fixed Point Results

First, we state the Nadler fixed point theorem in the class of M-metric spaces.

Theorem 1. *Let M-metric space (X,m) be M-complete and $F : X \to \mathcal{CB}^m(X)$ be a multivalued mapping. Suppose there exists $\lambda \in (0,1)$ such that*
$$\mathcal{H}_m(Fa,Fb) \leq \lambda m(a,b), \tag{1}$$

for all $a,b \in X$. Then F admits a fixed point.

Proof. Choose $q = \frac{1}{\sqrt{\lambda}}$ and $r = \sqrt{\lambda}$. Clearly, $q > 1$ and $r < 1$. Let $a_0 \in X$ be arbitrary and $a_1 \in Fa_0$. From Lemma 7, for $q = \frac{1}{\sqrt{\lambda}}$, there exists $a_2 \in Fa_1$ such that
$$m(a_1,a_2) \leq \frac{1}{\sqrt{\lambda}}\mathcal{H}_m(Fa_0,Fa_1). \tag{2}$$

As $\mathcal{H}_m(Fa_0,Fa_1) \leq \lambda m(a_0,a_1)$, so from (2) we have
$$m(a_1,a_2) \leq \frac{1}{\sqrt{\lambda}}\lambda m(a_0,a_1) = \sqrt{\lambda}m(a_0,a_1) = rm(a_0,a_1).$$

Now, from Lemma 7, there exists $a_3 \in Fa_2$ such that
$$m(a_2,a_3) \leq rm(a_1,a_2).$$

Continuing in this way, we get a sequence $\{a_k\}$ of points in X such that $a_{k+1} \in Fa_k$ and for $k \geq 1$,
$$m(a_k,a_{k+1}) \leq rm(a_{k-1},a_k), \tag{3}$$

that is,
$$m(a_k,a_{k+1}) \leq r^k m(a_0,a_1). \tag{4}$$

By Lemma 5, we have
$$\lim_{k \to \infty} m(a_k,a_{k+1}) = 0, \tag{5}$$

$$\lim_{k\to\infty} m(a_k, a_k) = 0, \tag{6}$$

and

$$\lim_{k,j\to\infty} m(a_k, a_j) = 0. \tag{7}$$

Also the sequence $\{a_k\}$ is M-Cauchy. Thus, M-completeness of X yields existence of $a \in X$ such that

$$\lim_{k\to\infty} (m(a_k, a) - m_{a_k a}) = 0.$$

Since $\lim_{k\to\infty} m(a_k, a_k) = 0$, we have

$$\lim_{k\to\infty} m(a_k, a) = 0. \tag{8}$$

From (1) and (8), we have

$$\lim_{k\to\infty} \mathcal{H}_m(Fa_k, Fa) = 0. \tag{9}$$

Now, since $a_{k+1} \in Fa_k$, $m(a_{k+1}, Fa) \leq \mathcal{H}_m(Fa_k, Fa)$. Taking limit as $k \to \infty$ and using (8), we get

$$\lim_{k\to\infty} m(a_{k+1}, Fa) = 0. \tag{10}$$

As $m_{a_{k+1} Fa} \leq m(a_{k+1}, Fa)$, so we have

$$\lim_{k\to\infty} m_{a_{k+1} Fa} = 0. \tag{11}$$

Using (m_4), we have

$$m(a, Fa) - \sup_{b \in Fa} m_{ab} \leq m(a, Fa) - m_{aFa}$$
$$\leq m(a, a_{k+1}) - m_{aa_{k+1}} + m(a_{k+1}, Fa) - m_{a_{k+1} Fa}.$$

Varying limit as $k \to \infty$ and using (8)–(11), we get

$$m(a, Fa) \leq \sup_{b \in Fa} m_{ab}. \tag{12}$$

Since $m_{ab} \leq m(a, b)$ for every $b \in Fa$, this implies that

$$m_{ab} - m(a, b) \leq 0.$$

Thus

$$\sup\{m_{ab} - m(a, b) : b \in Fa\} \leq 0,$$

that is,

$$\sup_{b \in Fa} m_{ab} - \inf_{b \in Fa} m(a, b) \leq 0.$$

This gives

$$\sup_{b \in Fa} m_{ab} \leq m(a, Fa). \tag{13}$$

From (12) and (13), we have

$$m(a, Fa) = \sup_{b \in Fa} m_{ab}.$$

Thus, by Lemma 6, $a \in \overline{Fa} = Fa$. □

Example 5. Let $X = [0,2]$ be endowed with m-metric $m(a,b) = |a-b| + \frac{a+b}{2}$. Then (X, m) is an M-complete M-metric space (as in Example 3). Let $F : X \to \mathcal{CB}^m(X)$ be a mapping defined as

$$F(a) = \left[0, \frac{1}{7}a^2\right] \text{ for all } a \in X.$$

We shall show that for $\lambda \in (0,1)$, $\mathcal{H}_m(Fa, Fb) \leq \lambda m(a,b)$, i.e., (1) holds for all $a,b \in X$. We have following three possible cases:

Case I: $a = b = p$. Then $Fa = [0, \frac{1}{7}p^2] = Fb$. Here, for $\lambda \geq \frac{2}{7}$,

$$\mathcal{H}_m(Fa, Fb) = \frac{1}{7}p^2 \leq \lambda p = \lambda m(p,p) = \lambda m(a,b).$$

Case II: $a < b$. Then $Fa = [0, \frac{1}{7}a^2]$, $Fb = [0, \frac{1}{7}b^2]$ and $Fa \subseteq Fb$. In this case,

$$\mathcal{H}_m(Fa, Fb) = \max\left\{\frac{1}{7}a^2, |\frac{1}{7}a^2 - \frac{1}{7}b^2| + \frac{\frac{1}{7}a^2 + \frac{1}{7}b^2}{2}\right\}.$$

Since $a < b$, $\frac{1}{7}a^2 < |\frac{1}{7}a^2 - \frac{1}{7}b^2| + \frac{\frac{1}{7}a^2 + \frac{1}{7}b^2}{2}$. So we get

$$\mathcal{H}_m(Fa, Fb) = |\frac{1}{7}a^2 - \frac{1}{7}b^2| + \frac{\frac{1}{7}a^2 + \frac{1}{7}b^2}{2}$$

and $m(a,b) = |a-b| + \frac{(a+b)}{2}$. Then one can see that

$$\mathcal{H}_m(Fa, Fb) = |\frac{1}{7}a^2 - \frac{1}{7}b^2| + \frac{\frac{1}{7}a^2 + \frac{1}{7}b^2}{2}$$
$$= \frac{1}{7}|(a-b)(a+b)| + \frac{1}{7}\frac{a^2+b^2}{2}$$
$$= \frac{1}{7}\left[|a-b|(a+b) + \frac{(a+b)^2 - 2ab}{2}\right]$$
$$\leq \frac{1}{7}\left[|a-b| + \frac{(a+b)}{2}\right](a+b)$$
$$= \frac{(a+b)}{7}m(a,b).$$

Case III: $a > b$. Then $Fa = [0, \frac{1}{7}a^2]$, $Fb = [0, \frac{1}{7}b^2]$ and $Fb \subseteq Fa$. In this case,

$$\mathcal{H}_m(Fa, Fb) = \max\left\{\frac{1}{7}b^2, |\frac{1}{7}a^2 - \frac{1}{7}b^2| + \frac{\frac{1}{7}a^2 + \frac{1}{7}b^2}{2}\right\}.$$

Since $b < a$, $\frac{1}{7}b^2 < |\frac{1}{7}a^2 - \frac{1}{7}b^2| + \frac{\frac{1}{7}a^2 + \frac{1}{7}b^2}{2}$. So, we get

$$\mathcal{H}_m(Fa, Fb) = |\frac{1}{7}a^2 - \frac{1}{7}b^2| + \frac{\frac{1}{7}a^2 + \frac{1}{7}b^2}{2}$$

and $m(a,b) = |a-b| + \frac{(a+b)}{2}$. Following Case II, one can easily show that

$$\mathcal{H}_m(Fa, Fb) \leq \frac{(a+b)}{7}m(a,b).$$

From above three cases, it is clear that (1) is satisfied for $\lambda \geq \frac{4}{7}$. Thus, all the required conditions of Theorem 1 are satisfied. Hence F admits a fixed point, which is $a = 0$.

Next, we present our fixed point result corresponding to multivalued Kannan contractions in M-metric spaces.

Theorem 2. *Let M-metric space (X, m) be M-complete and $F : X \to \mathcal{CB}^m(X)$ be a multivalued mapping. Suppose there exists $\lambda \in (0, \frac{1}{2})$ such that*

$$\mathcal{H}_m(Fa, Fb) \leq \lambda[m(a, Fa) + m(b, Fb)], \tag{14}$$

for all $a, b \in X$. Then F admits a fixed point in X.

Proof. Let $a_0 \in X$ be arbitrary. Fix an element $a_1 \in Fa_0$. We can now choose $a_2 \in Fa_1$ such that

$$m(a_1, a_2) = m(a_1, Fa_1) \leq \mathcal{H}_m(Fa_0, Fa_1).$$

Again, we can choose $a_3 \in Fa_2$ such that

$$m(a_2, a_3) \leq \mathcal{H}_m(Fa_1, Fa_2).$$

Continuing in this way, we get a sequence $\{a_k\}$ such that $a_{k+1} \in Fa_k$ with

$$m(a_k, a_{k+1}) \leq \mathcal{H}_m(Fa_{k-1}, Fa_k). \tag{15}$$

Using (14) in (15), we get

$$m(a_k, a_{k+1}) \leq \lambda[m(a_{k-1}, Fa_{k-1}) + m(a_k, Fa_k)]$$
$$\leq \lambda[m(a_{k-1}, a_k) + m(a_k, a_{k+1})].$$

Thus,

$$m(a_k, a_{k+1}) \leq \frac{\lambda}{1-\lambda} m(a_{k-1}, a_k).$$

Let $r = \frac{\lambda}{1-\lambda}$. Since $\lambda < \frac{1}{2}$, we have $r < 1$. So,

$$m(a_k, a_{k+1}) \leq r m(a_{k-1}, a_k). \tag{16}$$

Thus, from Lemma 5, we have

$$\lim_{k \to \infty} m(a_k, a_{k+1}) = 0, \tag{17}$$

$$\lim_{k \to \infty} m(a_k, a_k) = 0, \tag{18}$$

and

$$\lim_{k, j \to \infty} m(a_k, a_j) = 0. \tag{19}$$

Moreover, the sequence $\{a_k\}$ is a M-Cauchy. M-completeness of X yields existence of $a^* \in X$ such that

$$\lim_{k \to \infty} (m(a_k, a^*) - m_{a_k a^*}) = 0 \text{ and } \lim_{k \to \infty} (M_{a_k a^*} - m_{a_k a^*}) = 0.$$

Due to (18), we get

$$\lim_{k \to \infty} m(a_k, a^*) = 0 \text{ and } \lim_{k \to \infty} M_{a_k a^*} = 0.$$

Thus, we have

$$\lim_{k \to \infty} [M_{a_k a^*} + m_{a_k a^*}] = 0.$$

This implies that

$$m(a^*, a^*) = 0 \text{ and hence } m_{a^* Fa^*} = 0. \tag{20}$$

We shall show that $a^* \in Fa^*$. Since
$$m(a_{k+1}, Fa^*) \leq \mathcal{H}_m(Fa_k, Fa^*) \leq \lambda[m(a_k, Fa_k) + m(a^*, Fa^*)].$$

Taking limit as $k \to \infty$, we get
$$\lim_{k \to \infty} m(a_{k+1}, Fa^*) = 2\lambda m(a^*, Fa^*). \tag{21}$$

Suppose $m(a^*, Fa^*) > 0$, then we have
$$m(a^*, Fa^*) - m_{a^* Fa^*} \leq m(a^*, a_{k+1}) - m_{a^* a_{k+1}} + m(a_{k+1}, Fa^*) - m_{a_{k+1} Fa^*}.$$

Taking limit as $k \to \infty$ and using (21), we get $m(a^*, Fa^*) \leq 2\lambda m(a^*, Fa^*)$, which is a contradiction (as $2\lambda < 1$). So
$$m(a^*, Fa^*) = 0. \tag{22}$$

Also, using (20), we have
$$\sup_{b \in Fa} m_{a^* b} = \sup_{b \in Fa} \min\{m(a^*, a^*), m(b, b)\} = 0. \tag{23}$$

From (22) and (23), we get
$$m(a^*, Fa^*) = \sup_{b \in Fa} m_{a^* b}.$$

Thus, from Lemma 6, we get $a^* \in \overline{Fa^*} = Fa^*$. □

Example 6. Let $X = [0, 1]$ and $m : X \times X \to [0, \infty)$ be defined as
$$m(a, b) = \frac{a+b}{2}.$$

Then (X, m) is an M-complete M-metric space. Let $F : X \to \mathcal{CB}^m(X)$ be a mapping defined as
$$F(a) = \begin{cases} [0, a^2] & \text{if } a \in [0, \tfrac{1}{2}], \\ \left[\tfrac{a}{3}, \tfrac{a}{2}\right] & \text{if } a \in [\tfrac{1}{2}, 1]. \end{cases}$$

Then one can easily verify that there exists some λ in $(0, \tfrac{1}{2})$ such that
$$\mathcal{H}_m(Fa, Fb) \leq \lambda\big[m(a, Fa) + m(b, Fb)\big].$$

Thus F satisfies all the conditions in Theorem 2 and hence it has a fixed point (namely 0) in X.

Example 7. Let $X = [0, 1]$ be endowed with m-metric $m(x, y) = \tfrac{x+y}{2}$. Then (X, m) is an M-complete M-metric space. We define the mapping $F : X \to \mathcal{CB}^m(X)$ as
$$F(a) = \begin{cases} \{\tfrac{1}{5}\} & \text{if } a = 0, \\ \left[\dfrac{a}{8(1+a^2)}, \dfrac{a}{4(1+a^2)}\right] & \text{if } a > 0. \end{cases}$$

For $a = 0$ and $b = \frac{1}{10}$, there does not exist any λ in $(0, \frac{1}{2})$ such that

$$\mathcal{H}_m(F(0), F(\tfrac{1}{10})) \leq \lambda [m(0, F(0)) + m(\tfrac{1}{10}, F(\tfrac{1}{10}))].$$

Thus F does not satisfy (14) in Theorem 2. Evidently, F has no fixed point in X.

5. Homotopy Results in M-Metric Spaces

The following result is required in the sequel while proving a homotopy result in M-metric spaces.

Proposition 3. *Let $F: X \to \mathcal{CB}^m(X)$ be a multivalued mapping satisfying (1) for all a, b in M-metric space (X, m). If $c \in Fc$ for some $c \in X$, then $m(a, a) = 0$ for $a \in Fc$.*

Proof. Let $c \in Fc$. Then $m(c, Fc) = \sup_{b \in Fc} m_{c,b} = \sup_{b \in Fc} m_{bb}$. Also

$$\mathcal{H}_m(Fc, Fc) = \delta_m(Fc, Fc) = \sup_{b \in Fc} m_{bb}.$$

Assume that $m(c, c) > 0$. We have

$$\sup_{b \in Fc} m_{bb} = \mathcal{H}_m(Fc, Fc) \leq \lambda m(c, c),$$

that is,

$$\sup_{b \in Fc} m_{bb} \leq \lambda m(c, c).$$

Since $c \in Fc$, it is a contradiction. So $m(a, a) = 0$ for every $a \in Fc$. □

Theorem 3. *Let $\mathcal{O}$ (resp. $\mathcal{C}$) be an open (resp. closed) subset in an M-complete M-metric space (X, m) such that $\mathcal{O} \subset \mathcal{C}$. Let $\mathcal{G} : \mathcal{C} \times [\mu, \nu] \to \mathcal{CB}^m(X)$ be a mapping satisfying the following conditions:*

(a) *$a \notin \mathcal{G}(a, t)$ for all $a \in \mathcal{C} \setminus \mathcal{O}$ and each $t \in [\mu, \nu]$;*
(b) *there exists $\lambda \in (0, 1)$ such that for every $t \in [\mu, \nu]$ and all $a, b \in \mathcal{C}$ we have*

$$\mathcal{H}_m(\mathcal{G}(a, t), \mathcal{G}(b, t)) \leq \lambda m(a, b);$$

(c) *there exists a continuous mapping $\psi : [\mu, \nu] \to \mathbb{R}$ satisfying*

$$\mathcal{H}_m(\mathcal{G}(a, t), \mathcal{G}(a, s)) \leq \lambda |\psi(t) - \psi(s)|;$$

(d) *if $c \in \mathcal{G}(c, t)$ then $\mathcal{G}(c, t) = \{c\}$.*

If $\mathcal{G}(., t_1)$ admits a fixed point in $\mathcal{C}$ for at least one $t_1 \in [\mu, \nu]$, then $\mathcal{G}(., t)$ admits a fixed point in $\mathcal{O}$ for all $t \in [\mu, \nu]$. Moreover, the fixed point of $\mathcal{G}(., t)$ is unique for any fixed $t \in [\mu, \nu]$.

Proof. Consider, the set

$$\mathcal{W} = \{t \in [\mu, \nu] | a \in \mathcal{G}(a, t) \text{ for some } a \in \mathcal{O}\}.$$

Then $\mathcal{W}$ is nonempty, because $\mathcal{G}(., t_1)$ has a fixed point in $\mathcal{C}$ for at least one $t_1 \in [\mu, \nu]$, that is, there exists $a \in \mathcal{C}$ such that $a \in \mathcal{G}(a, t_1)$ and as (a) holds, we have $a \in \mathcal{O}$.

We will show that $\mathcal{W}$ is both closed and open in $[\mu, \nu]$. First, we show that it is open.

Let $t_0 \in \mathcal{W}$ and $a_0 \in \mathcal{O}$ with $a_0 \in \mathcal{G}(a_0, t_0)$. As $\mathcal{O}$ is open subset of X, $B_m(a_0, r) \subseteq \mathcal{O}$ for some $r > 0$. Let $\varepsilon = r + m_{aa_0} - \lambda(r + m_{aa_0}) > 0$. As ψ is continuous on $[\mu, \nu]$, there exists $\delta > 0$ such that

$$|\psi(t) - \psi(t_0)| < \epsilon, \text{ for all } t \in S_\delta(t_0),$$

where $S_\delta(t_0) = (t_0 - \delta, t_0 + \delta)$.

Since $a_0 \in \mathcal{G}(a_0, t_0)$, by Proposition 3, $m(c,c) = 0$ for every $c \in \mathcal{G}(a_0, t_0)$. Keeping this fact in view, we have
$$m_{pc} = 0, \text{ for every } p \in X. \tag{24}$$

Now, using (iii) of Proposition 2 and (24), we have

$$\begin{aligned}
m(\mathcal{G}(a,t), a_0) &= \mathcal{H}_m(\mathcal{G}(a,t), \mathcal{G}(a_0, t_0)) \\
&\leq \mathcal{H}_m(\mathcal{G}(a,t), \mathcal{G}(a, t_0)) + \mathcal{H}_m(\mathcal{G}(a, t_0), \mathcal{G}(a_0, t_0)) \\
&\quad - \inf_{p \in \mathcal{G}(a,t)} \inf_{q \in \mathcal{G}(a, t_0)} m_{pq} - \inf_{q \in \mathcal{G}(a, t_0)} \inf_{c \in \mathcal{G}(a_0, t_0)} m_{qc} + \sup_{p \in \mathcal{G}(a,t)} \sup_{c \in \mathcal{G}(a_0, t_0)} m_{pc} \\
&\leq \mathcal{H}_m(\mathcal{G}(a,t), \mathcal{G}(a, t_0)) + \mathcal{H}_m(\mathcal{G}(a, t_0), \mathcal{G}(a_0, t_0)) \\
&\leq \lambda |\psi(t) - \psi(t_0)| + \lambda m(a, a_0) \\
&\leq \lambda \varepsilon + \lambda(m_{aa_0} + r) \\
&= \lambda(r + m_{aa_0} - \lambda(r + m_{aa_0})) + \lambda(m_{aa_0} + r) \\
&\leq r + m_{aa_0} - \lambda(r + m_{aa_0}) + \lambda(m_{aa_0} + r) \\
&\leq r + m_{aa_0}.
\end{aligned}$$

Thus for each fixed $t \in S_\delta(t_0)$, $\mathcal{G}(.,t) : \overline{B_m(a_0, r)} \to \mathcal{CB}^m(\overline{B_m(a_0, r)})$ satisfies all the hypotheses of Theorem 1 and so $\mathcal{G}(.,t)$ admits a fixed point in $\overline{B_m(a_0, r)} \subseteq \mathcal{C}$. But this fixed point must be in $\mathcal{O}$ to satisfy (a). Therefore, $S_\delta(t_0) \subseteq \mathcal{W}$ and hence $\mathcal{W}$ is open in $[\mu, \nu]$.

Next, we show that $\mathcal{W}$ is closed in $[\mu, \nu]$. Let $\{t_k\}$ be a convergent sequence in $\mathcal{W}$ to some $s \in [\mu, \nu]$. We need to show that $s \in \mathcal{W}$.

The definition of the set $\mathcal{W}$ implies that for all $k \in \mathbb{N} \setminus \{0\}$, there exists $a_k \in \mathcal{O}$ with $a_k \in \mathcal{G}(a_k, t_k)$. Then using (d), (iii) of Proposition 2 and the outcome of Proposition 3, we have

$$\begin{aligned}
m(a_k, a_j) &= \mathcal{H}_m(\mathcal{G}(a_k, t_k), \mathcal{G}(a_j, t_j)) \\
&\leq \mathcal{H}_m(\mathcal{G}(a_k, t_k), \mathcal{G}(a_k, t_j)) + \mathcal{H}_m(\mathcal{G}(a_k, t_j), \mathcal{G}(a_j, t_j)) \\
&\leq \lambda |\psi(t_k) - \psi(t_j)| + \lambda m(a_k, a_j).
\end{aligned}$$

This gives us
$$m(a_k, a_j) \leq \frac{\lambda}{1-\lambda} |\psi(t_k) - \psi(t_j)|.$$

Since ψ is continuous and $\{t_k\}$ converges to s, varying $k, j \to \infty$ in the above inequality, we get
$$\lim_{k,j \to \infty} m(a_k, a_j) = 0.$$

As $m_{a_k a_j} \leq m(a_k, a_j)$, so
$$\lim_{k,j \to \infty} m_{a_k a_j} = 0.$$

Also $\lim_{k \to \infty} m(a_k, a_k) = 0 = \lim_{k \to \infty} m(a_j, a_j)$.
Therefore
$$\lim_{k,j \to \infty} (m(a_k, a_j) - m_{a_k a_j}) = 0 \text{ and } \lim_{k,j \to \infty} (M_{a_k a_j} - m_{a_k a_j}) = 0.$$

Thus $\{a_k\}$ is an M-Cauchy sequence. Using (iii) of Definition 3, there exists $a^* \in X$ such that
$$\lim_{k \to \infty} (m(a_k, a^*) - m_{a_k a^*}) = 0 \text{ and } \lim_{k \to \infty} (M_{a_k, a^*} - m_{a_k a^*}) = 0.$$

But $\lim_{k\to\infty} m(a_k, a_k) = 0$, so

$$\lim_{k\to\infty} m(a_k, a^*) = 0 \text{ and } \lim_{k\to\infty} M_{a_k a^*} = 0.$$

Thus, we get $m(a^*, a^*) = 0$. We shall prove $a^* \in \mathcal{G}(a^*, t^*)$. We have

$$\begin{aligned} m(a_k, \mathcal{G}(a^*, t^*)) &\leq \mathcal{H}_m(\mathcal{G}(a_k, t_k), \mathcal{G}(a^*, t^*)) \\ &\leq \mathcal{H}_m(\mathcal{G}(a_k, t_k), \mathcal{G}(a_k, t^*)) + \mathcal{H}_m(\mathcal{G}(a_k, t^*), \mathcal{G}(a^*, t^*)) \\ &\leq \lambda |\psi(a_k) - \psi(t^*)| + \lambda m(a_k, a^*). \end{aligned}$$

Varying $k \to \infty$ in above inequality, we get

$$\lim_{k\to\infty} m(a_k, \mathcal{G}(a^*, t^*)) = 0.$$

Hence

$$m(a^*, \mathcal{G}(a^*, t^*)) = 0. \tag{25}$$

Since $m(a^*, a^*) = 0$, we have

$$\sup_{b \in \mathcal{G}(a^*, t^*)} m_{a^* b} = \sup_{b \in \mathcal{G}(a^*, t^*)} \min\{m(a^*, a^*), m(b, b)\} = 0. \tag{26}$$

From (25) and (26), we get

$$m(a^*, \mathcal{G}(a^*, t^*)) = \sup_{b \in \mathcal{G}(a^*, t^*)} m_{a^* b}.$$

Therefore, from Lemma 6, we have $a^* \in \mathcal{G}(a^*, t^*)$. Thus $a^* \in \mathcal{O}$. Hence $t^* \in \mathcal{W}$ and $\mathcal{W}$ is closed in $[\mu, \nu]$.

As $[\mu, \nu]$ is connected and $\mathcal{W}$ is both open and closed in it, so $\mathcal{W} = [\mu, \nu]$. Thus $\mathcal{G}(., t)$ admits a fixed point in $\mathcal{O}$ for all $t \in [\mu, \nu]$.

For uniqueness, fix $t \in [\mu, \nu]$, then there exists $a \in \mathcal{O}$ such that $a \in \mathcal{G}(a, t)$. Suppose b is another fixed point of $\mathcal{G}(b, t)$, then from (d) we have

$$m(a, b) = \mathcal{H}_m(\mathcal{G}(a, t), \mathcal{G}(b, t)) \leq \lambda m(a, b),$$

a contradiction. Thus, the fixed point of $\mathcal{G}(., t)$ is unique for any $t \in [\mu, \nu]$. □

Author Contributions: All authors contributed equally to this paper. All authors have read and approved the final manuscript.

Funding: This research received no external funding.

Acknowledgments: The fifth author would like to thank Prince Sultan University for funding this work through research group Nonlinear Analysis Methods in Applied Mathematics (NAMAM) group number RG-DES-2017-01-17.

Conflicts of Interest: The authors declare no conflict of interest.

References

1. Nadler, S.B. Multi-valued contraction mappings. *Pac. J. Math.* **1969**, *30*, 475–488. [CrossRef]
2. Reich, S. Fixed points of contractive functions. *Boll. dell'Unione Mat. Ital.* **1972**, *5*, 17–31.
3. Reich, S. Approximate selections, best approximations, fixed points, and invariant sets. *J. Math. Anal. Appl.* **1978**, *62*, 104–113. [CrossRef]
4. Matthews, S.G. Partial metric topology. *Ann. N. Y. Acad. Sci.* **1994**, *728*, 183–197. [CrossRef]

5. Aydi, H.; Barakat, M.; Felhi, A.; Isik, H. On phi-contraction type couplings in partial metric spaces. *J. Math. Anal.* **2017**, *8*, 78–89.
6. Ciric, L.; Samet, B.; Aydi, H.; Vetro, C. Common fixed points of generalized contractions on partial metric spaces and an application. *Appl. Math. Comput.* **2011**, *218*, 2398–2406.
7. Abodayeh, K.; Mliaki, N.; Abdeljawad, T.; Shatanawi, W. Relation Between Partial Metric Spaces and M-Metric Spaces, Caristi Kirk's Theorem in M-Metric Type Spaces. *J. Math. Anal.* **2016**, *7*, 1–12.
8. Ameer, E.; Aydi, H.; Arshad, M.; Alsamir, H.; Noorani, M.S. Hybrid multivalued type contraction mappings in α_K-complete partial b-metric spaces and applications. *Symmetry* **2019**, *11*, 86. [CrossRef]
9. Aydi, H.; Felhi, A.; Karapinar, E.; Sahmim, S. A Nadler-type fixed point theorem in dislocated spaces and applications. *Miscolc Math. Notes* **2018**, *19*, 111–124. [CrossRef]
10. Karapinar, E.; Shatanawi, W.; Tas, K. Fixed point theorem on partial metric spaces involving rational expressions. *Miskolc Math. Notes* **2013**, *14*, 135–142. [CrossRef]
11. Shatanawi, W.; Pitea, A. Some coupled fixed point theorems in quasi-partial metric spaces. *Fixed Point Theory Appl.* **2013**, *2013*, 153. [CrossRef]
12. Karapinar, E.; Agarwal, R.P.; Aydi, H. Interpolative Reich-Rus-Ciric type contractions on partial metric spaces. *Mathematics* **2018**, *6*, 256. [CrossRef]
13. Shatanawi, W.; Postolache, M. Coincidence and fixed point results for generalized weak contractions in the sense of Berinde on partial metric spaces. *Fixed Point Theory Appl.* **2013**, *2013*, 54. [CrossRef]
14. Karapinar, E.; Shatanawi, W. On Weakly (C, ψ, ϕ)-Contractive Mappings in Ordered Partial Metric Spaces. *Abstr. Appl. Anal.* **2012**, *2012*, 495892. [CrossRef]
15. Aydi, H.; Karapinar, E.; Shatanawi, W. Coupled fixed point results for (ψ, φ)-weakly contractive condition in ordered partial metric spaces. *Comput. Math. Appl.* **2011**, *62*, 4449–4460. [CrossRef]
16. Shatanawi, W.; Nashine, H.K.; Tahat, N. Generalization of some coupled fixed point results on partial metric spaces. *Int. J. Math. Math. Sci.* **2012**, *2012*, 686801. [CrossRef]
17. Asadi, M.; Karapinar, E.; Salimi, P. New extension of p-metric spaces with fixed-point results on M-metric spaces. *J. Inequal. Appl.* **2014**, *2014*, 18. [CrossRef]
18. Souayah, N.; Mlaiki, N.; Mrad, M. The G_M−Contraction Principle for Mappings on M−Metric Spaces Endowed With a Graph and Fixed Point Theorems. *IEEE Access* **2018**, *6*, 25178–25184. [CrossRef]
19. Aydi, H.; Abbas, M.; Vetro, C. Partial Hausdorff metric and Nadler's fixed point theorem on partial metric spaces. *Topol. Appl.* **2012**, *159*, 3234–3242. [CrossRef]
20. Aydi, H.; Abbas, M.; Vetro, C. Common Fixed points for multivalued generalized contractions on partial metric spaces, RACSAM—Revista de la Real Academia de Ciencias Exactas. *Fisicas y Naturales Serie A Matematicas* **2014**, *108*, 483–501.

© 2019 by the authors. Licensee MDPI, Basel, Switzerland. This article is an open access article distributed under the terms and conditions of the Creative Commons Attribution (CC BY) license (http://creativecommons.org/licenses/by/4.0/).

Article

Common Fixed Point under Nonlinear Contractions on Quasi Metric Spaces

Wasfi Shatanawi [1,2,*] and Kamaleldin Abodayeh [1]

1. Department of Mathematics and General Sciences, Prince Sultan University, Riyadh 11942, Saudi Arabia; kamal@psu.edu.sa
2. Department of Medical Research, China Medical University Hospital, China Medical University, Taichung 40402, Taiwan
* Correspondence: wshatanawi@psu.edu.sa or wshatanawi@yahoo.com

Received: 16 March 2019; Accepted: 9 May 2019; Published: 20 May 2019

Abstract: We introduce in this article the notion of (ψ, ϕ)−quasi contraction for a pair of functions on a quasi-metric space. We also investigate the existence and uniqueness of the fixed point for a couple functions under that contraction.

Keywords: quasi metric space; altering distance function; (ψ, ϕ)−quasi contraction.

1. Introduction and Preliminary

Fixed point has been considered by many researchers since it was established by Banach [1] in 1992. The generalizations of the theory were considered by many researchers on various metric spaces (see, for example, [2–7]). Quasi-metric space was one of the interesting examples that were considered since it was introduced by Wilson [8] in 1931. We may suggest the following articles to the reader [8–20].

Definition 1. *[8] Let χ be a non-empty set and $\rho : \chi \times \chi \to [0, \infty)$ be a given function that satisfies the following conditions:*

(1) $\rho(\alpha, \beta) = 0$ *if and only if* $\alpha = \beta$.
(2) $\rho(\alpha, \beta) \leq \rho(\alpha, \gamma) + \rho(\gamma, \beta)$ *for all* $\alpha, \beta, \gamma \in \chi$.

Then, ρ is called a quasi-metric on χ and the pair (χ, ρ) is called a quasi-metric space.

Example 1. *Consider the set $\chi = [0, 1]$ and define the function $\rho : \chi \times \chi \to [0, \infty)$ such that*

$$\rho(\alpha, \beta) = \begin{cases} \alpha^2 - \beta^2 & \text{if } \alpha \geq \beta \\ 1 & \text{Otherwise}. \end{cases}$$

Then, (χ, ρ) is a quasi-metric space. To prove this, we need to verify the two conditions of Definition 1.

Condition 1. If $\alpha = \beta$, then it is clear that $\rho(\alpha, \beta) = 0$. On the other hand, if $\rho(\alpha, \beta) = 0$ then we have $0 = \alpha^2 - \beta^2 = (\alpha - \beta)(\alpha + \beta)$. Since $\alpha, \beta \in [0, 1]$, we have $\alpha = \beta$.

Condition 2. Let $\alpha, \beta, \gamma \in \chi$. Then, we have three cases:

  Case I If $\alpha > \beta$ and $\beta > \gamma$, then $\alpha > \gamma$ and hence $\rho(\alpha, \beta) + \rho(\beta, \gamma) = (\alpha^2 - \beta^2) + (\beta^2 - \gamma^2) = \alpha^2 - \gamma^2 = \rho(\alpha, \gamma)$.
  Case II If $\alpha > \beta$ and $\gamma > \beta$, then we have $\rho(\alpha, \beta) + \rho(\beta, \gamma) = (\alpha^2 - \beta^2) + 1 > \rho(\alpha, \gamma)$. This is because $\rho(\alpha, \gamma) \leq 1$ for all $\alpha, \gamma \in [0, 1]$.
  Case III If $\beta > \alpha$, then using the same reason as in Case II, we have $\rho(\alpha, \beta) + \rho(\beta, \gamma) = 1 + \rho(\beta, \gamma) > \rho(\alpha, \gamma)$.

Therefore, (χ, ρ) is a quasi-metric space. It is clear that (χ, ρ) is not a metric space since $\rho(\alpha, \beta) \neq \rho(\beta, \alpha)$, for all $\alpha \neq \beta$.

Now, we introduce the definitions of convergence and Cauchy of such a sequence in quasi-metric spaces:

Definition 2. *[12,13] Let (χ, ρ) be a quasi-metric space. A sequence (α_n) in χ converges to the element $\alpha \in \chi$ if and only if*
$$\lim_{n \to \infty} \rho(\alpha_n, \alpha) = \lim_{n \to \infty} \rho(\alpha, \alpha_n) = 0.$$

Definition 3. *[12,13] Let (χ, ρ) be a quasi-metric space. A sequence (α_n) in the space χ is said to be a Cauchy sequence if and if, for $\epsilon > 0$, there exists a positive integer $N = N(\epsilon)$ such that $\rho(\alpha_n, \alpha_m) < \epsilon$ for all $m, n > N$.*

Moreover, if every Cauchy sequence in the quasi-metric space χ is convergent, then (χ, ρ) is said to be complete.

The next notion was given by Khan et al. [21].

Definition 4. *[21] A self function ψ on $[0, \infty)$ is called an altering distance function if the following properties hold:*

(1) *ψ is non-decreasing and continuous.*
(2) *$\psi(e) = 0$ if and only if $e = 0$.*

2. Main Result

Definition 5. *Let (χ, ρ) be a quasi-metric space and S_1, S_2 be two self-mappings on χ. Then, the pair (S_1, S_2) is said to be (ψ, ϕ)–quasi contraction if there exist two alternating distance functions ψ and ϕ such that, for all $e, w \in \chi$, we have*
$$\psi(\rho(S_1 e, S_2 w)) \leq \psi(M_1(e, w)) - \phi(M_1(e, w))$$
and
$$\psi(\rho(S_2 e, S_1 w)) \leq \psi(M_2(e, w)) - \phi(M_2(e, w))$$
where
$$M_1(e, w) = \max\left\{ \rho(w, S_2 w) \frac{1 + \rho(e, S_1 e)}{1 + \rho(e, w)}, \rho(e, S_1 e), \rho(w, S_2 w) \right\}$$
and
$$M_2(e, w) = \max\left\{ \rho(w, S_1 w) \frac{1 + \rho(e, S_2 e)}{1 + \rho(e, w)}, \rho(e, S_2 e), \rho(w, S_1 w) \right\}.$$

Now, we prove our first result:

Theorem 1. *Let (χ, ρ) be a complete quasi-metric space. Let ψ and ϕ be alternating distance functions and S_1, S_2 be two self-mappings on χ such that the pair (S_1, S_2) is (ψ, ϕ)–quasi contraction. Then, S_1 and S_2 have a unique common fixed point.*

Proof. We start the proof of the result by taking an element $\tau_0 \in \chi$. We construct a sequence (τ_n) in χ in the following way: $\tau_{2n+1} = S_1 \tau_{2n}$ and $\tau_{2n+2} = S_2 \tau_{2n+1}$ for all $n \geq 0$.

It is clear that if there exists $s \in \mathbb{N}$ with $\tau_{2s} = \tau_{2s+1}$, then τ_{2s} is a fixed point of S_1. Since the pair (S_1, S_2) is (ψ, ϕ)–quasi contraction, we have

$$\begin{aligned}
&\psi(\rho(\tau_{2s+1}, \tau_{2s+2})) \\
={}& \psi(\rho(S_1 \tau_{2s}, S_2 \tau_{2s+1})) \\
\leq{}& \psi(M_1(\tau_{2s}, \tau_{2s+1})) - \phi(M_1(\tau_{2s}, \tau_{2s+1})) \\
={}& \psi\left(\max\{\rho(\tau_{2s+1}, \tau_{2s+2}), \rho(\tau_{2s}, \tau_{2s+1})\}\right) \\
& -\phi\left(\max\{\rho(\tau_{2s+1}, \tau_{2s+2}), \rho(\tau_{2s}, \tau_{2s+1})\}\right) \\
={}& \psi(\rho(\tau_{2s+1}, \tau_{2s+2})) - \phi(\rho(\tau_{2s+1}, \tau_{2s+2})).
\end{aligned}$$

From the above inequality, we deduce that $\psi(\rho(\tau_{2s+1}, \tau_{2s+2})) = 0$. Since ψ is an alternating function, we conclude that τ_{2s} is a fixed point of S_1 and S_2. Thus, τ_{2s} is a common fixed point of S_1 and S_2.

Using similar arguments as above, we may show that, if there exists $s \in \mathbb{N}$ such that $\tau_{2s+1} = \tau_{2s+2}$, then τ_{2s+1} is a common fixed point of S_1 and S_2.

Now, we may assume that $\tau_n \neq \tau_{n+1}$ for all $n \in \mathbb{N}$.

In view of (ψ, ϕ)–quasi contraction of the pair (S_1, S_2), we deduce that

$$\begin{aligned}
&\psi(\rho(\tau_{2n+1}, \tau_{2n+2})) \\
={}& \psi(\rho(S_1 \tau_{2n}, S_2 \tau_{2n+1})) \\
\leq{}& \psi(M_1(\tau_{2n}, \tau_{2n+1})) - \phi(M_1(\tau_{2n}, \tau_{2n+1})) \\
={}& \psi\left(\max\left\{\rho(\tau_{2n+1}, \tau_{2n+2})\frac{1+\rho(\tau_{2n}, \tau_{2n+1})}{1+\rho(\tau_{2n}, \tau_{2n+1})}, \rho(\tau_{2n}, \tau_{2n+1}), \rho(\tau_{2n+1}, \tau_{2n+2})\right\}\right) \\
& -\phi\left(\max\left\{\rho(\tau_{2n+1}, \tau_{2n+2})\frac{1+\rho(\tau_{2n}, \tau_{2n+1})}{1+\rho(\tau_{2n}, \tau_{2n+1})}, \rho(\tau_{2n}, \tau_{2n+1}), \rho(\tau_{2n+1}, \tau_{2n+2})\right\}\right) \\
={}& \psi(\max\{\rho(\tau_{2n}, \tau_{2n+1}), \rho(\tau_{2n+1}, \tau_{2n+2})\}) - \phi(\max\{\rho(\tau_{2n}, \tau_{2n+1}), \rho(\tau_{2n+1}, \tau_{2n+2})\}). \quad (1)
\end{aligned}$$

Assume that

$$\max\left\{\psi(\rho(\tau_{2n}, \tau_{2n+1})), \psi(\rho(\tau_{2n+1}, \tau_{2n+2}))\right\} = \psi(\rho(\tau_{2n+1}, \tau_{2n+2})).$$

Then, Equation (1) implies

$$\psi(\rho(\tau_{2n+1}, \tau_{2n+2})) \leq \psi(\rho(\tau_{2n+1}, \tau_{2n+2})) - \phi(\rho(\tau_{2n+1}, \tau_{2n+2}))$$

a contradiction. Thus,

$$\max\left\{\psi(\rho(\tau_{2n}, \tau_{2n+1})), \psi(\rho(\tau_{2n+1}, \tau_{2n+2}))\right\} = \psi(\rho(\tau_{2n}, \tau_{2n+1})).$$

Therefore, Equation (1) yields

$$\psi(\rho(\tau_{2n+1}, \tau_{2n+2})) \leq \psi(\rho(\tau_{2n}, \tau_{2n+1})) - \phi(\rho(\tau_{2n}, \tau_{2n+1})). \quad (2)$$

On the other hand, we have

$$\begin{aligned}
&\psi(\rho(\tau_{2n}, \tau_{2n+1})) \\
=\ & \psi(\rho(S_2\tau_{2n-1}, S_1\tau_{2n})) \\
\leq\ & \psi(M_2(\tau_{2n-1}, \tau_{2n})) - \phi(M_2(\tau_{2n-1}, \tau_{2n})) \\
=\ & \psi\left(\max\left\{\rho(\tau_{2n}, \tau_{2n+1})\frac{1+\rho(\tau_{2n-1}, \tau_{2n})}{1+\rho(\tau_{2n-1}, \tau_{2n})}, \rho(\tau_{2n-1}, \tau_{2n}), \rho(\tau_{2n}, \tau_{2n+1})\right\}\right) \\
& - \phi\left(\max\left\{\rho(\tau_{2n}, \tau_{2n+1})\frac{1+\rho(\tau_{2n-1}, \tau_{2n})}{1+\rho(\tau_{2n-1}, \tau_{2n})}, \rho(\tau_{2n-1}, \tau_{2n}), \rho(\tau_{2n}, \tau_{2n+1})\right\}\right) \\
=\ & \psi(\max\{\rho(\tau_{2n-1}, \tau_{2n}), \rho(\tau_{2n}, \tau_{2n+1})\}) - \phi(\max\{\rho(\tau_{2n-1}, \tau_{2n}), \rho(\tau_{2n}, \tau_{2n+1})\}). \quad (3)
\end{aligned}$$

From the last inequality, we get

$$\max\{\rho(\tau_{2n-1}, \tau_{2n}), \rho(\tau_{2n}, \tau_{2n+1})\} = \rho(\tau_{2n-1}, \tau_{2n}),$$

and hence

$$\psi(\rho(\tau_{2n}, \tau_{2n+1})) \leq \psi(\rho(\tau_{2n-1}, \tau_{2n})) - \phi(\rho(\tau_{2n-1}, \tau_{2n})). \quad (4)$$

Combining Equations (2) and (4), we conclude that

$$\psi(\rho(\tau_n, \tau_{n+1})) \leq \psi(\rho(\tau_{n-1}, \tau_n)) - \phi(\rho(\tau_{n-1}, \tau_n)) < \psi(\rho(\tau_{n-1}, \tau_n)) \quad (5)$$

holds for all $n \in \mathbb{N}$.

From Equation (5), we conclude that $\{\rho(\tau_{n-1}, \tau_n) : n = 1, 2, \ldots\}$ is a decreasing sequence. There exists $s \geq 0$ such that

$$\lim_{n \to +\infty} \rho(\tau_{n-1}, \tau_n) = s.$$

By allowing n tends to $+\infty$ in Equation (5), we conclude that $s = 0$ and hence

$$\lim_{n \to +\infty} \rho(\tau_{n-1}, \tau_n) = 0. \quad (6)$$

Now, we prove that

$$\lim_{n,m \to +\infty} \rho(\tau_n, \tau_m) = 0.$$

For two large integer numbers n and m with $m > n$, we discuss the following cases:

Case 1: $n = 2l + 1$ and $m = 2r + 2$ for some $l, r \in \mathbb{N}$; that is, n is odd and m is even. By the (ϕ, ψ)-contraction of the pair (S_1, S_2), we have

$$\begin{aligned}
&\psi(\rho(\tau_n, \tau_m)) \\
=\ & \psi(\rho(\tau_{2l+1}, \tau_{2r+2})) \\
=\ & \psi(\rho(S_1\tau_{2l}, S_2\tau_{2r+1})) \\
\leq\ & \psi(M_1(\tau_{2l}, \tau_{2r+1})) - \phi(M_1(\tau_{2l}, \tau_{2r+1})) \\
=\ & \psi\left(\max\left\{\rho(\tau_{2r+1}, \tau_{2r+2})\frac{1+\rho(\tau_{2l}, \tau_{2l+1})}{1+\rho(\tau_{2l}, \tau_{2r+1})}, \rho(\tau_{2l}, \tau_{2l+1}), \rho(\tau_{2r+1}, \tau_{2r+2})\right\}\right) \\
& - \phi\left(\max\left\{\rho(\tau_{2r+1}, \tau_{2r+2})\frac{1+\rho(\tau_{2l}, \tau_{2l+1})}{1+\rho(\tau_{2l}, \tau_{2r+1})}, \rho(\tau_{2l}, \tau_{2l+1}), \rho(\tau_{2r+1}, \tau_{2r+2})\right\}\right)
\end{aligned}$$

$$\quad (7)$$

$$\begin{aligned}
&\leq \psi\left(\max\left\{\rho(\tau_{2r+1},\tau_{2r+2})(1+\rho(\tau_{2l},\tau_{2l+1})),\rho(\tau_{2l},\tau_{2l+1}),\rho(\tau_{2r+1},\tau_{2r+2})\right\}\right) \\
&\quad -\phi(\rho(\tau_{2l},\tau_{2l+1})) \\
&\leq \psi\left(\max\left\{\rho(\tau_{2r+1},\tau_{2r+2})(1+\rho(\tau_{2l},\tau_{2l+1})),\rho(\tau_{2l},\tau_{2l+1}),\rho(\tau_{2r+1},\tau_{2r+2})\right\}\right) \\
&\leq \psi(\rho(\tau_{2l+1},\tau_{2l+2})(1+\rho(\tau_{2l},\tau_{2l+1}))) \\
&= \psi(\rho(\tau_n,\tau_{n+1})(1+\rho(\tau_{n-1},\tau_n))) \\
&\leq \psi(\rho(\tau_{n-1},\tau_n)(1+\rho(\tau_{n-1},\tau_n))).
\end{aligned}$$

In view of Equation (7) and the nondecreasing property of the function ψ, we conclude that

$$\rho(\tau_n,\tau_m) \leq \rho(\tau_{n-1},\tau_n)(1+\rho(\tau_{n-1},\tau_n))$$

Case 2: $n=2l$ and $m=2r+2$ for some $l,r \in \mathbb{N}$; that is, n and m are both even. Here, we have

$$\begin{aligned}
\rho(\tau_n,\tau_m) = \rho(\tau_{2l},\tau_{2r+2}) &\leq \rho(\tau_{2l},\tau_{2l+1}) + \rho(\tau_{2l+1},\tau_{2r+2}) \\
&= \rho(\tau_n,\tau_{n+1}) + \rho(\tau_{n+1},\tau_m).
\end{aligned}$$

From Case 1, we get

$$\begin{aligned}
\rho(\tau_n,\tau_m) &\leq \rho(\tau_n,\tau_{n+1}) + \rho(\tau_n,\tau_{n+1})(1+\rho(\tau_n,\tau_{n+1})) \\
&\leq \rho(\tau_{n-1},\tau_n) + \rho(\tau_{n-1},\tau_n)(1+\rho(\tau_{n-1},\tau_n)).
\end{aligned}$$

Case 3: $n=2l$ and $m=2r+3$ for some $l,r \in \mathbb{N}$; that is, n is an even number and m is an odd number. Here, we have

$$\begin{aligned}
\rho(\tau_n,\tau_m) &= \rho(\tau_{2l},\tau_{2r+3}) \\
&\leq \rho(\tau_{2l},\tau_{2l+1}) + \rho(\tau_{2l+1},\tau_{2r+2}) + \rho(\tau_{2r+2},\tau_{2r+3}) \\
&= \rho(\tau_n,\tau_{n+1}) + \rho(\tau_{n+1},\tau_{m-1}) + \rho(\tau_{m-1},\tau_m).
\end{aligned}$$

From Case 1, we get

$$\begin{aligned}
\rho(\tau_n,\tau_m) &\leq \rho(\tau_n,\tau_{n+1}) + \rho(\tau_{n+1},\tau_{m-1}) + \rho(\tau_{m-1},\tau_m) \\
&\leq \rho(\tau_n,\tau_{n+1}) + \rho(\tau_n,\tau_{n+1})(1+\rho(\tau_n,\tau_{n+1})) + \rho(\tau_{m-1},\tau_m) \\
&\leq 2\rho(\tau_{n-1},\tau_n) + \rho(\tau_{n-1},\tau_n)(1+\rho(\tau_{n-1},\tau_n)).
\end{aligned}$$

Case 4: $n=2l+1$ and $m=2r+3$ for some $l,r \in \mathbb{N}$; that is, n and m are both odd. Here, we have

$$\begin{aligned}
\rho(\tau_n,\tau_m) &= \rho(\tau_{2l+1},\tau_{2r+3}) \\
&\leq \rho(\tau_{2l+1},\tau_{2r+2}) + \rho(\tau_{2r+2},\tau_{2r+3}) \\
&= \rho(\tau_n,\tau_{m-1}) + \rho(\tau_{m-1},\tau_m).
\end{aligned}$$

Case 1 implies that

$$\begin{aligned}
\rho(\tau_n,\tau_m) &\leq \rho(\tau_{n-1},\tau_n)(1+\rho(\tau_{n-1},\tau_n)) + \rho(\tau_{m-1},\tau_m) \\
&\leq \rho(\tau_{n-1},\tau_n)(1+\rho(\tau_{n-1},\tau_n)) + \rho(\tau_{n-1},\tau_n).
\end{aligned}$$

By summing all cases together, we conclude that

$$\rho(\tau_n, \tau_m) \leq 2\rho(\tau_{n-1}, \tau_n) + \rho(\tau_n, \tau_{n+1})(1 + \rho(\tau_{n-1}, \tau_n)) \quad (8)$$

holds for all $n, m \in \mathbb{N}$.

Letting $n, m \to +\infty$ in (8), we have

$$\lim_{n,m \to +\infty} \rho(\tau_n, \tau_m) = 0.$$

Thus, (τ_n) is a Cauchy sequence in χ. In view of the competence of the space χ, we find $a \in \chi$ such that $\tau_n \to a$ as n tends to $+\infty$.

For $s \in \mathbb{N}$, we have

$$\begin{aligned}
&\psi(\rho(\tau_{2s+1}, S_2 a)) \\
&= \psi(\rho(S_1 \tau_{2s}, S_2 a)) \\
&\leq \psi(M_1(\tau_{2s}, a)) - \phi(M_1(\tau_{2s}, a)) \\
&= \psi\left(\max\left\{\rho(a, S_2 a)\frac{1 + \rho(\tau_{2s}, \tau_{2s+1})}{1 + \rho(\tau_{2s}, a)}, \rho(\tau_{2s}, \tau_{2s+1}), \rho(a, S_2 a)\right\}\right) \\
&\quad - \phi\left(\max\left\{\rho(a, S_2 a)\frac{1 + \rho(\tau_{2s}, \tau_{2s+1})}{1 + \rho(\tau_{2s}, a)}, \rho(\tau_{2s}, \tau_{2s+1}), \rho(a, S_2 a)\right\}\right).
\end{aligned}$$

Allowing $s \to +\infty$ in above inequality, we get

$$\psi(\rho(a, S_2 a)) \leq \psi(\rho(a, S_2 a)) - \phi(\rho(a, S_2 a)).$$

The above inequality is correct only if $\phi(\rho(a, S_2 a)) = 0$ and thus $S_2 a = a$. Using similar arguments as above, we may figure out $S_1 a = a$. Thus, a is a common fixed point of S_1 and S_2.

Now, assume that $S_1 w_1 = S_2 w_1 = w_1$ and $S_1 w_2 = S_2 w_2 = w_2$. In view of (ψ, ϕ)−contraction of the pair (S_1, S_2), we have

$$\psi(\rho(w_1, w_2)) = \psi(\rho(S_1 w_1, S_2 w_2)) \leq 0.$$

Thus, $\psi(\rho(w_1, w_2)) = 0$. Therefore, $w_1 = w_2$. Thus, the common fixed point of S_1 and S_2 is unique. □

By taking

$$\max\left\{\rho(w, S_2 w)\frac{1 + \rho(e, S_1 e)}{1 + \rho(e, w)}, \rho(e, S_1 e), \rho(w, S_2 w)\right\} = \rho(w, S_2 w)\frac{1 + \rho(e, S_1 e)}{1 + \rho(e, w)}$$

and

$$\max\left\{\rho(w, S_1 w)\frac{1 + \rho(e, S_2 e)}{1 + \rho(e, w)}, \rho(e, S_2 e), \rho(w, S_1 w)\right\} = \rho(w, S_1 w)\frac{1 + \rho(e, S_2 e)}{1 + \rho(e, w)}$$

in Definition 5. Then, the following result holds:

Corollary 1. *Let (χ, ρ) be a complete quasi-metric space and $S_1, S_2 : \chi \to \chi$ be two mappings. Let ψ and ϕ be two altering distance functions such that*

$$\psi(\rho(S_1 e, S_2 w)) \leq \psi\left(\rho(w, S_2 w)\frac{1 + \rho(e, S_1 e)}{1 + \rho(e, w)}\right) - \phi\left(\rho(w, S_2 w)\frac{1 + \rho(e, S_1 e)}{1 + \rho(e, w)}\right),$$

and

$$\psi(\rho(S_2 e, S_1 w)) \leq \psi\left(\rho(w, S_1 w)\frac{1 + \rho(e, S_2 e)}{1 + \rho(e, w)}\right) - \phi\left(\rho(w, S_1 w)\frac{1 + \rho(e, S_2 e)}{1 + \rho(e, w)}\right).$$

Then, S_1 and S_2 have a unique common fixed point.

If we define ψ and ϕ on the interval $[0, +\infty)$ such that $\psi(\tau) = \tau$ and $\phi(\tau) = (1-a)\tau$ where $a \in [0,1)$ in Theorem 1, we formulate the following result.

Corollary 2. *Let (χ, ρ) be a complete quasi-metric space and $S_1, S_2 : \chi \to \chi$ be two mappings. Let $a \in [0,1)$ such that*

$$\psi(\rho(S_1 e, S_2 w)) \leq a \max\left\{\rho(w, S_2 w)\frac{1+\rho(e, S_1 e)}{1+\rho(e, w)}, \rho(e, S_1 e), \rho(w, S_2 w)\right\},$$

and

$$\psi(\rho(S_2 e, S_1 w)) \leq a \max\left\{\rho(w, S_1 w)\frac{1+\rho(e, S_1 2e)}{1+\rho(e, w)}, \rho(e, S_1 2e), \rho(w, S_1 w)\right\}.$$

Then, S_1 and S_2 have a unique common fixed point.

In addition, if we assume $S_1 = S_2$ in Theorem 1, Corollary 1, and Corollary 2, then the following results hold.

Corollary 3. *Let (χ, ρ) be a complete quasi-metric space and S_1 be a self-mapping on χ. Assume ψ and ϕ are two altering distance functions such that*

$$\begin{aligned}\psi(\rho(S_1 e, S_1 w)) &\leq \psi\left(\left\{\rho(w, S_1 w)\frac{1+\rho(e, S_1 e)}{1+\rho(e, w)}, \rho(e, S_1 e), \rho(w, S_1 w)\right\}\right) \\ &\quad -\phi\left(\left\{\rho(w, S_1 w)\frac{1+\rho(e, S_1 e)}{1+\rho(e, w)}, \rho(e, S_1 e), \rho(w, S_1 w)\right\}\right).\end{aligned}$$

Then, S_1 has a unique fixed point.

Corollary 4. *Let (χ, ρ) be a complete quasi-metric space and $S_1 : \chi \to \chi$ be a mapping. Let ψ and ϕ be two altering distance functions such that*

$$\psi(\rho(S_1 e, S_1 w)) \leq \psi\left(\max\left\{\rho(w, S_1 w)\frac{1+\rho(e, S_1 e)}{1+\rho(e, w)}\right\}\right) - \phi\left(\max\left\{\rho(w, S_1 w)\frac{1+\rho(e, S_1 e)}{1+\rho(e, w)}\right\}\right).$$

Then, S_1 has a unique fixed point.

Corollary 5. *Let (χ, ρ) be a complete quasi-metric space and $S_1 : \chi \to \chi$ be a mapping. Let $a \in [0,1)$ such that*

$$\psi(\rho(S_1 e, S_1 w)) \leq a \max\left\{\rho(w, S_1 w)\frac{1+\rho(e, S_1 e)}{1+\rho(e, w)}, \rho(e, S_1 e), \rho(w, S_1 w)\right\}.$$

Then, S_1 has a unique fixed point.

The following example shows the validate of our results:

Example 2. *On the space $\chi = [0,1]$, define the quasi-metric via*

$$\rho(\alpha, \beta) = \begin{cases} 0, & \text{if } \alpha = \beta; \\ \max\{\alpha, \beta\}, & \text{if } \alpha \neq \beta, \end{cases}$$

In addition, on $\chi = [0,1]$, define the mappings S_1 and S_2 via $S_1\tau = \frac{1}{2}\sin^2\tau$ and $S_2\tau = \frac{1}{2}\tau^2$. Take the following altering functions $\psi(\alpha) = \frac{\alpha}{1+\alpha}$ and $\phi(\alpha) = \frac{\alpha}{5+5\alpha}$. Then,

1. ρ induces complete quasi-metric on χ.
2. (S_1, S_2) is (ψ, ϕ) − quasi contraction.

Proof. The proof of Part (1) is clear. To verify Part (2), given $(e,w) \in [0,1] \times [0,1]$ with $e \neq w$. Without loss of generality, we may assume that $e > w$. Then,

$$M_1(e,w) = \left\{ \rho\left(w, \frac{1}{2}w^2\right)\left(\frac{1+\rho(e,\frac{1}{2}\sin^2 e)}{1+\rho(e,w)}\right), \rho(e, \frac{1}{2}\sin^2 e), \rho(w, \frac{1}{2}w^2) \right\}$$
$$= e.$$

Thus,

$$\psi(\rho(S_1 e, S_2 w)) = \psi\left(\rho\left(\frac{1}{2}\sin^2 e, \frac{1}{2}w^2\right)\right)$$
$$= \frac{\max\{\frac{1}{2}\sin^2 e, \frac{1}{2}w^2\}}{1 + \max\{\frac{1}{2}\sin^2 e, \frac{1}{2}w^2\}}$$
$$\leq \frac{\frac{1}{2}e}{1 + \frac{1}{2}e}$$
$$= \frac{e}{2+e}$$
$$\leq \left(\frac{4}{5}\right)\left(\frac{e}{1+e}\right)$$
$$= \frac{e}{1+e} - \frac{e}{5+5e}$$
$$= \psi(M_1(e,w)) - \phi(M_1(e,w)).$$

Using similar arguments as for the above method, we can prove that

$$\psi(\rho(S_2 e, S_1 w)) = \psi(M_2(e,w)) - \phi(M_2(e,w)).$$

Thus, (S_1, S_2) is (ψ, ϕ)−quasi contraction. Thus, by Theorem 1, we deduce that S_1 and S_2 have a unique common fixed point. □

Author Contributions: Both authors contributed equally and significantly in writing this article. Both authors read and approved the final manuscript.

Funding: The authors thanks Prince Sultan University for supporting this paper through the research group Nonlinear Analysis Methods in Applied Mathematics (NAMAM), group number RG-DES-2017-01-17.

Conflicts of Interest: The authors declare no conflict of interest.

References

1. Banach, S. Sur Les opérations dans les ensembles abstraits et leur application aux équations intégrals. *Fund. Math.* **1922**, *3*, 133–181. [CrossRef]
2. Shatanawi, W.; Postolache, M. Common fixed point results of mappings for nonlinear contraction of cyclic form in ordered metric spaces. *Fixed Point Theory Appl.* **2013**, *60*, 1–13.
3. Shatanawi, W.; Postolache, M. Coincidence and fixed point results for generalized weak contractions in the sense of Berinde on partial metric spaces. *Fixed Point Theory Appl.* **2013**, *2013*, 54. [CrossRef]

4. Abdeljawad, T.; Karapınar, E.; Tas, K. Existence and uniqueness of a common fixed point on partial metric spaces. *Appl. Math. Lett.* **2011**, *24*, 1900–1904.
5. Shatanawi, W. Some fixed point results for a generalized ψ-weak contraction mappings in orbitally metric spaces. *Chaos Solitons Fractals* **2012**, *45*, 520–526. [CrossRef]
6. Abodayeh, K.; Shatanawi, W.; Bataihah, A.; Ansari, A.H. Some fixed point and common fixed point results through Ω-distance under nonlinear contractions. *Gazi J. Sci.* **2017**, *30*, 293–302.
7. Shatanawi, W. Fixed and common fixed point theorems in frame of quasi-metric spaces under contraction condition based on ultra distance functions. *Nonlinear Anal. Modell. Control* **2018**, *23*, 724–748
8. Wilson, W.A. On quasi-metric spaces. *Am. J. Math.* **1931**, *53*, 675–684. [CrossRef]
9. Alqahtani, B.; Fulga, A.; Karapınar, E. Fixed point results on Δ-symmetric quasi-metric space via simulation function with an application to Ulam stability. *Mathematics* **2018**, *6*, 208. [CrossRef]
10. Aydi, H.; Felhi, A.; Karapınar, E.; Alojail, F.A. Fixed points on quasi-metric spaces via simulation functions and consequences. *J. Math. Anal.* **2018**, *9*, 10–24.
11. Gregoria, V.; Romaguerab, S. Fixed point theorems for fuzzy mappings in quasi-metric spaces. *Fuzzy Sets Syst.* **2000**, *115*, 477–483. [CrossRef]
12. Aydi, H.; Jellali, M.; Karapınar, E. On fixed point results for $\alpha-$implicit contractions in quasi-metric spaces and consequences. *Nonlinear Anal. Model. Control* **2016**, *21*, 40–56. [CrossRef]
13. Jleli, M.; Samet, B. Remarks on G -metric spaces and fixed point theorems. *Fixed Point Theory Appl.* **2012**, *2012*, 210. [CrossRef]
14. Aydi, H. $\alpha-$implicit contractive pair of mappings on quasi $b-$metric spaces and an application to integral equations. *J. Nonlinear Convex Anal.* **2016**, *17*, 2417–2433.
15. Afshari, H.; Kalantari, S.; Aydi, H. Fixed point results for generalized $\alpha - \psi -$ Suzuki$-$contractions in quasi$-b-$metric-like spaces. *Asian-Eur. J. Math.* **2018**, *11*, 1850012. [CrossRef]
16. Felhi, A.; Sahmim, S.; Aydi, H. Ulam-Hyers stability and well-posedness of fixed point problems for $\alpha - \lambda -$contractions on quasi $b-$metric spaces. *Fixed Point Theory Appl.* **2016**, *2016*, 1. [CrossRef]
17. Bilgili, N.; Karapinar, E.; Samet, B. Generalized quasi-metric $\alpha - \psi$ contractive mappings in quasi-metric spaces and related fixed-point theorems. *J. Inequal. Appl.* **2014**, *36*, 1–15. [CrossRef]
18. Abodayeh, K.; Shatanawi, W.; Turkoglu, D. Some fixed point theorems in quasi-metric spaces under quasi weak contractions. *Glob. J. Pure Appl. Math.* **2016**, *12*, 4771–4780.
19. Shatanawi, W.; Pitea, A. Some coupled fixed point theorems in quasi-partial metric spaces. *Fixed Point Theory Appl.* **2013**, *153*, 1–15. [CrossRef]
20. Shatanawi, W.; Noorani, M.S.; Alsamir, H.; Bataihah, A. Fixed and common fixed point theorems in partially ordered quasi-metric spaces. *J. Math. Comput. Sci.* **2016**, *16*, 516–528. [CrossRef]
21. Khan, M.S.; Swaleh, M.; Sessa, S. Fixed point theorems by altering distances between the points. *Bull. Aust. Math. Soc.* **1984**, *30*, 1–9. [CrossRef]

© 2019 by the authors. Licensee MDPI, Basel, Switzerland. This article is an open access article distributed under the terms and conditions of the Creative Commons Attribution (CC BY) license (http://creativecommons.org/licenses/by/4.0/).

Article
On Pata–Suzuki-Type Contractions

Obaid Alqahtani [1], Venigalla Madhulatha Himabindu [2] and Erdal Karapınar [3],*

[1] Department of Mathematics, King Saud University, Riyadh 11451, Saudi Arabia
[2] Department of Mathematics, Koneru Lakshmaiah Educational Foundation, Vaddeswaram, Guntur, Andhra Pradesh 522502, India
[3] Department of Medical Research, China Medical University Hospital, China Medical University, Taichung 40402, Taiwan
* Correspondence: karapinar@mail.cmuh.org.tw

Received: 24 June 2019; Accepted: 30 July 2019; Published: 8 August 2019

Abstract: In this paper, we aim to obtain fixed-point results by merging the interesting fixed-point theorem of Pata and Suzuki in the framework of complete metric space and to extend these results by involving admissible mapping. After introducing two new contractions, we investigate the existence of a (common) fixed point in these new settings. In addition, we shall consider an integral equation as an application of obtained results.

Keywords: pata type contraction; Suzuki type contraction; C-condition; orbital admissible mapping

MSC: 54H25; 47H10; 54E50

1. Introduction and Preliminaries

For the solution of several differential/fractional/integral equations, fixed-point theory plays a significant role. In such investigations, usually well-known Banach fixed-point theorems are sufficient to provide the desired results. In the case of the inadequacy, the researcher in the fixed-point theory proposes some extension of the Banach contraction principle. Among them, we recall one of the significant theorems given by Popescu [1] inspired from the notion of C-condition defined by Suzuki [2].

Definition 1 (See [3])**.** *Let T be a self-mapping on a metric space (X, d). It is called C-condition if*

$$\frac{1}{2}d(\varkappa, T\varkappa) \leq d(\varkappa, y) \text{ implies that } d(T\varkappa, Ty) \leq d(\varkappa, y), \forall \varkappa, y \in X.$$

Indeed, by using the notion of C-condition, Suzuki [2] extended the famous Edelstein Theorem. More precisely, For a self-mapping T on a compact metric space (X, d), if T is C-condition and the inequality $d(T\varkappa, Ty) < d(\varkappa, y)$, for all $\varkappa \neq y$, then T possesses a unique fixed point.

Popescu [1] considered Bogin-type fixed-point theorem involving the notion of C-condition in a complete metric space as follows:

Theorem 1. *Let a self-mapping T on a complete metric space (X, d) satisfy the following condition:*

$$\frac{1}{2}d(\varkappa, T\varkappa) \leq d(\varkappa, y) \tag{1}$$

implies

$$d(T\varkappa, Ty) \leq ad(\varkappa, y) + b[d(\varkappa, T\varkappa) + d(y, Ty)] + c[d(\varkappa, Ty) + d(y, T\varkappa)] \tag{2}$$

where $a \geq 0$, $b > 0$, $c > 0$ and $a + 2b + 2c = 1$. Then T has a unique fixed point.

Another outstanding generalization of Banach mapping principle was given by Pata [4]. Before giving the result of Pata [4], we fix some notations:

For an arbitrary point $\varkappa_0$ in a complete metric space (X,d), we shall consider a functional

$$\|\varkappa\| = d(\varkappa, \varkappa_0), \forall \varkappa \in X,$$

that will be called "the zero of X". In addition, $\psi : [0,1] \to [0,\infty)$ will be a fixed increasing function that is continuous at zero, with $\psi(0) = 0$.

Theorem 2 (See [4]). *Let T be a self-mapping on a metric space (X,d). Suppose that $\beta \in [0,\alpha]$ $\Lambda \geq 0$ and $\alpha \geq 1$ are fixed constants. A self-mapping T possesses a unique fixed point if*

$$d(T\varkappa, Ty) \leq (1-\varepsilon)d(\varkappa,y) + \Lambda(\varepsilon)^\alpha \psi(\varepsilon)\left[1 + \|\varkappa\| + \|y\|\right]^\beta,$$

holds for all $\varkappa, y \in X$ and for every $\varepsilon \in [0,1]$.

This theorem has been extended, modified, and generalized by several authors, e.g., [5–16].

The main goal of this paper is to introduce new contractions that are inspired from the results of Suzuki [2], Popescu [1], and Pata [4]. More precisely, our new contraction not only merges these two successful generalization Banach contractions, but also extends the structure by involving α-admissible mappings in it. After that, we aim to investigate the existence and uniqueness of this new contraction in the context of complete metric spaces.

For this purpose, we recall some basic notions and results from recent literature.

Definition 2 ([17]). *Let $X \neq \emptyset$ and $\alpha : X \times X \to [0,\infty)$ be an auxiliary function. A self-mapping T on X is called α-orbital admissible if*

$$\alpha(\varkappa, T\varkappa) \geq 1 \text{ implies that } \alpha(T\varkappa, T^2\varkappa) \geq 1, \text{ for any } \varkappa \in X.$$

Lemma 1 (See[18]). *Let $\{p_n\}$ be a sequence on a metric space (X,d). Suppose that the sequence $\{d(p_{n+1}, p_n)\}$ is nonincreasing with*

$$\lim_{n\to\infty} d(p_{n+1}, p_n) = 0,$$

If $\{p_n\}$ is not a Cauchy sequence then there exists a $\delta > 0$ and two strictly increasing sequences $\{m_k\}$ and $\{n_k\}$ in $\mathbb{N}$ such that the following sequences tend to δ:

$$d(p_{m_k}, p_{n_k}), d(p_{m_k}, p_{n_k+1}), d(p_{m_k-1}, p_{n_k}), d(p_{m_k-1}, p_{n_k+1}), d(p_{m_k+1}, p_{n_k+1}),$$

as $k \to \infty$.

2. Main Results

We start with the definition of the α-Pata–Suzuki contraction:

Definition 3. *Let (X,d) be a metric space and let $\Lambda \geq 0$, $\alpha \geq 1$ and $\beta \in [0,\alpha]$ be fixed constants. A self-mapping T, defined on X, is called α-Pata–Suzuki contraction if for every $\varepsilon \in [0,1]$ and all $x,y \in X$, satisfies the following condition*

(i) *T is an α-orbital admissible mapping*
(ii)

$$\frac{1}{2}d(x, Tx) \leq d(x,y)$$

implies

$$\alpha(x, Tx)\alpha(y, Ty)d(Tx, Ty) \leq P(x,y)$$

where

$$P(x,y) = (1-\varepsilon) \max \left\{ d(x,y), d(x,Tx), d(y,Ty), \tfrac{1}{2}[d(x,Ty) + d(y,Tx)] \right\}$$
$$+ \Lambda(\varepsilon)^\alpha \psi(\varepsilon) [1 + \|x\| + \|y\| + \|Tx\| + \|Ty\|]^\beta.$$

This is the first main result of this paper.

Theorem 3. *Let (X,d) be a metric space and T be a self-mapping on X. If*

(i) *T on X is α-Pata–Suzuki contraction;*
(ii) *there exists $x_0 \in X$ such that $\alpha(x_0, Tx_0) \geq 1$;*
(iii) *if $\{x_n\}$ is a sequence in X such that $\alpha(x_n, x_{n+1}) \geq 1$ for all n and $x_n \to x$ as $n \to \infty$, then $\alpha(x_n, x) \geq 1$ for all n, we have $\alpha(x, Tx) \geq 1$;*
(iv) *$\alpha(x^*, Tx^*) \geq 1$ for all $x^* \in Fix(T)$, where $Fix(T) := \{x \in X : Tx = x\}$, then T has a fixed point $z \in X$.*

Proof. Due to assumptions of the theorem, there is $x_0 \in X$ so that $\alpha(x_0, Tx_0) \geq 1$. In addition, we set $\|x\| = d(x, x_0), \forall x \in X$. Since T is an α-orbital admissible mapping, we have

$$\alpha(Tx_0, T^2 x_0) \geq 1.$$

and iteratively, we have

$$\alpha(T^n x_0, T^{n+1} x_0) \geq 1 \text{ for each } n \in \mathbb{N}. \qquad (3)$$

Starting at this point x_0 we shall construct an iterative sequence $\{x_n\}$ by $x_n = T^n x_0$ for $n = 1, 2, 3, \cdots$. Here, we assume that consequent terms are distinct. Indeed, if there exists $k_0 \in \mathbb{N}$ such that

$$T_0^k x_0 = x_{k_0} = x_{k_0+1} = T^{k_0+1} x_0 = T(T^k x_0) = T(x_{k_0}),$$

then, x_{k_0} forms a fixed point. To avoid from the trivial case, we suppose that

$$x_n \neq x_{n+1} \text{ for all } n = 1, 2, 3, \cdots.$$

To prove that the sequence $\{d(x_n, x_{n+1})\}$ is decreasing, suppose on the contrary that

$$d(x_n, x_{n+1}) = \max\{d(x_n, x_{n+1}), d(x_n, x_{n-1})\}.$$

Since $\tfrac{1}{2} d(x_{n-1}, x_n) \leq d(x_{n-1}, x_n)$ and since T is a α-Pata–Suzuki contraction, we find that

$$d(x_n, x_{n+1}) = d(Tx_{n-1}, Tx_n)$$
$$\leq \alpha(x_{n-1}, Tx_{n-1}) \alpha(x_n, Tx_n) d(Tx_{n-1}, Tx_n)$$
$$\leq (1-\varepsilon) \max \left\{ d(x_{n-1}, x_n), d(x_{n-1}, x_n), d(x_n, x_{n+1}), \tfrac{1}{2}[d(x_n, x_n) + d(x_{n-1}, x_{n+1})] \right\}$$
$$+ \Lambda(\varepsilon)^\alpha \psi(\varepsilon) [1 + \|x_{n-1}\| + \|x_n\| + \|Tx_{n-1}\| + \|Tx_n\|]^\beta$$
$$\leq (1-\varepsilon) d(x_n, x_{n+1}) + K(\varepsilon)^\alpha \psi(\varepsilon),$$

for some $K > 0$. It follows that $d(x_n, x_{n+1}) = 0$ which is a contradiction. Hence, $\{d(x_n, x_{n+1})\}$ is a decreasing sequence, thus tending to some non-negative real number, say, d^*.

As a next step, we shall show that the sequence $\{\|x_n\|\}$ is bounded. For simplicity, let $C_n = \|x_n\|$, and hence, we claim that the sequence $\{C_n\}$ is bounded.

Since the sequence $\{d(x_n, x_{n+1})\}$ is decreasing, from the triangle inequality, we find that

$$C_n = d(x_n, x_0) \leq d(x_n, x_{n+1}) + d(Tx_n, Tx_0) + C_1$$
$$\leq 2C_1 + d(Tx_n, Tx_0).$$

We assert that

$$\frac{1}{2}d(x_n, x_{n+1}) \leq d(x_n, x_0) \text{ or } \frac{1}{2}d(x_{n-1}, x_n) \leq d(x_{n-1}, x_0).$$

Suppose, on contrary that

$$\frac{1}{2}d(x_n, x_{n+1}) > d(x_n, x_0) \text{ and } \frac{1}{2}d(x_{n-1}, x_n) > d(x_{n-1}, x_0).$$

In this case, we derive that

$$\begin{aligned}
d(x_{n-1}, x_n) &\leq d(x_{n-1}, x_0) + d(x_0, x_n) \\
&< \frac{1}{2}[d(x_{n-1}, x_n) + d(x_n, x_{n+1})] \\
&\leq d(x_{n-1}, x_n),
\end{aligned}$$

is a contradiction. Hence, our assertion is held, i.e.,

$$\frac{1}{2}d(x_n, x_{n+1}) \leq d(x_n, x_0) \text{ or } \frac{1}{2}d(x_{n-1}, x_n) \leq d(x_{n-1}, x_0).$$

Also, on account of (3), we have

$$\alpha(x_n, Tx_n)\alpha(x_0, Tx_0) \geq 1.$$

Regarding T is α-Pata–Suzuki contraction, we get

$$d(Tx_n, Tx_0) \leq \alpha(x_n, Tx_n)\alpha(x_0, Tx_0)d(Tx_n, Tx_0)$$
$$\leq (1-\varepsilon)\max\left\{ d(x_n, x_0), \ d(x_0, x_1), \ d(x_n, x_{n+1}), \ \tfrac{1}{2}[d(x_n, x_1) + d(x_0, x_{n+1})] \right\}$$
$$+ \Lambda(\varepsilon)^\alpha \psi(\varepsilon)[1 + \|x_n\| + \|x_0\| + \|x_{n+1}\| + \|x_1\|]^\beta$$
$$\leq (1-\varepsilon)\max\{C_n, C_1, C_1 + C_n\} + \Lambda(\varepsilon)^\alpha \psi(\varepsilon)[1 + C_n + C_1 + C_1 + C_n]^\beta$$
$$\leq (1-\varepsilon)(C_1 + C_n) + \Lambda(\varepsilon)^\alpha \psi(\varepsilon)[1 + 2C_n + 2C_1]^\beta.$$

Consequently, we derive from the above inequality that

$$C_n = d(x_n, x_0) \leq d(x_n, x_{n+1}) + d(fx_n, fx_0) + C_1$$
$$\leq 2C_1 + (1-\varepsilon)(C_1 + C_n) + a(\varepsilon)^\alpha \psi(\varepsilon).$$

A simple calculation yields that

$$\varepsilon C_n \leq a(\varepsilon)^\alpha \psi(\varepsilon) + b,$$

for some constants $a, b > 0$. By verbatim of the proof of ([18], Lemma 1.5) it follows that the sequence $\{C_n\}$ is bounded.

In what follows we prove that $d^* = 0$ by employing the fact that $\{C_n\}$ is bounded. Indeed, we have that

$$\begin{aligned}
d(x_{n+1}, x_n) &= d(Tx_n, Tx_{n-1}) \\
&\leq \alpha(x_{n-1}, Tx_{n-1})\alpha(x_n, Tx_n) d(Tx_n, Tx_{n-1}) \\
&\leq (1-\varepsilon) d(x_n, x_{n-1}) + \Lambda(\varepsilon)^\alpha \psi(\varepsilon) \left[1 + \|x_n\| + \|x_{n-1}\| + \|x_n\| + \|x_{n+1}\|\right]^\beta \\
&\leq (1-\varepsilon) d(x_n, x_{n-1}) + \Lambda(\varepsilon)^\alpha \psi(\varepsilon) \left[1 + 2\|x_n\| + \|x_{n-1}\| + \|x_{n+1}\|\right]^\beta \\
&\leq (1-\varepsilon) d(x_n, x_{n-1}) + K(\varepsilon)^\alpha \psi(\varepsilon),
\end{aligned}$$

for some $K > 0$. As $n \to \infty$ in the inequality above, it follows that $d^* = 0$.

As a next step, we shall indicate that $\{x_n\}$ is a Cauchy sequence by using the method of *Reductio ad Absurdum*. Assume, on the contrary, that the sequence $\{x_n\}$ is not Cauchy. Accordingly, regarding on Lemma 1, there exists $\delta > 0$ and two increasing sequences $\{m_k\}$ and $\{n_k\}$, with $n_k > m_k > k$ such that the sequences $d(x_{m_k}, x_{n_k}), d(x_{m_k}, x_{n_{k+1}}), d(x_{m_{k-1}}, x_{n_k}), d(x_{m_{k-1}}, x_{n_{k+1}}), d(x_{m_{k+1}}, x_{n_k})$ tends to δ as $n \to \infty$.

We claim that $\frac{1}{2} d(x_{m_{k-1}}, x_{m_k}) \leq d(x_{m_{k-1}}, x_{n_k})$. Indeed, if the inequality above is not held, that is, if $\frac{1}{2} d(x_{m_{k-1}}, x_{m_k}) > d(x_{m_{k-1}}, x_{n_k})$ then we get a contradiction. More precisely, by letting $k \to \infty$ in the previous inequality, we get $\delta \leq 0$, a contradiction.

Hence, our claim is valid, i.e., $\frac{1}{2} d(x_{m_{k-1}}, x_{m_k}) \leq d(x_{m_{k-1}}, x_{n_k})$. Notice also that $\alpha(x_{m_{k-1}}, f(x_{m_{k-1}})) \alpha(x_{n_k}, f x_{n_k}) \geq 1 \, \forall k \geq N$. Since T is α-Pata–Suzuki contraction, we deduce that

$$\begin{aligned}
d(x_{m_k}, x_{n_{k+1}}) &= d(Tx_{m_{k-1}}, Tx_{n_k}) \\
&\leq \alpha(x_{m_{k-1}}, T(x_{m_{k-1}})) \alpha(x_{n_k}, T, x_{n_k}) d(Tx_{m_{k-1}}, Tx_{n_k}) \\
&\leq (1-\varepsilon) \max \left\{ \begin{array}{c} d(x_{m_{k-1}}, x_{n_k}), d(x_{m_{k-1}}, x_{m_k}), d(x_{n_k}, x_{n_{k+1}}), \\ \frac{1}{2} [d(x_{n_k}, x_{m_k}) + d(x_{m_{k-1}}, x_{n_{k+1}})] \end{array} \right\} \\
&\quad + \Lambda(\varepsilon)^\alpha \psi(\varepsilon) \left[1 + \|x_{m_{k-1}}\| + \|x_{n_k}\| + \|x_{m_k}\| + \|x_{n_{k+1}}\|\right]^\beta \\
&\leq (1-\varepsilon) \max \left\{ \begin{array}{c} d(x_{m_{k-1}}, x_{n_k}), d(x_{m_{k-1}}, x_{m_k}), d(x_{n_k}, x_{n_{k+1}}), \\ \frac{1}{2} [d(x_{n_k}, x_{m_k}) + d(x_{m_{k-1}}, x_{n_{k+1}})] \end{array} \right\} \\
&\quad + K(\varepsilon)^\alpha \psi(\varepsilon),
\end{aligned}$$

where $K > 0$. By letting $k \to \infty$ in the obtained inequality above, we get that $\delta = 0$, a contradiction.

Hence, $\{x_n\}$ is a Cauchy sequence. Since X is complete, there exists $z^* \in X$ such that $x_n \to z^*$ and by (v) and $\alpha(z^*, Tz^*) \geq 1$.

Now, we shall prove that $z^* = Tz^*$. Suppose, on the contrary, that $z^* \neq Tz^*$. For this purpose, we need to prove the claim: For each $n \geq 1$, at least one of the following assertions holds.

$$\frac{1}{2} d(x_{n-1}, x_n) \leq d(x_{n-1}, z^*) \text{ or } \frac{1}{2} d(x_n, x_{n+1}) \leq d(x_n, z^*).$$

Again, we use the method of Reductio ad Absurdum and assume it does not hold, i.e.,

$$\frac{1}{2} d(x_{n-1}, x_n) > d(x_{n-1}, z^*) \text{ and } \frac{1}{2} d(x_n, x_{n+1}) > d(x_n, z^*),$$

for some $n \geq 1$. Then, keeping in mind that $\{d(x_n, x_{n+1})\}$ is a decreasing sequence, the triangle inequality infers

$$\begin{aligned} d(x_{n-1}, x_n) &\leq d(x_{n-1}, z^*) + d(z^*, x_n) \\ &< \tfrac{1}{2}[d(x_{n-1}, x_n) + d(x_n, x_{n+1})] \\ &< d(x_{n-1}, x_n), \end{aligned}$$

which is a contradiction, and so the claim holds.

Due to the assumption (v) and the observation (3), we have

$$\alpha(x_n, Tx_n)\alpha(z^*, Tz^*) \geq 1, \text{ holds for all } n \in N.$$

Taking $\tfrac{1}{2}d(x_n, Tx_n) \leq d(x_n, z^*)$ into account, the assumption (i) yields that

$$d(Tx_n, Tz^*) \leq (1-\varepsilon) \max\left\{ d(x_n, z^*),\ d(z^*, Tz^*),\ d(x_n, x_{n+1}),\ \tfrac{1}{2}[d(x_n, Tz^*) + d(z^*, Tx_n)] \right\}$$

$$+ \Lambda(\varepsilon)^\alpha \psi(\varepsilon) \left[1 + \|x_n\| + \|z^*\| + \|Tz^*\| + \|Tx_n\|\right]^\beta$$

$$= (1-\varepsilon) \max\left\{ d(x_n, z^*),\ d(z^*, Tz^*),\ d(x_n, x_{n+1}),\ \tfrac{1}{2}[d(x_n, Tz^*) + d(z^*, Tx_n)] \right\}$$

$$+ K(\varepsilon)^\alpha \psi(\varepsilon),$$

for some $K > 0$. By letting $n \to \infty$ in the inequality above, we find that

$$\begin{aligned} d(z^*, fz_1) &\leq (1-\varepsilon) \max\left\{0, d(z^*, Tz^*), 0, \tfrac{d(z^*, Tz^*)}{2}\right\} + K(\varepsilon)^\alpha \psi(\varepsilon) \\ &< (1-\varepsilon) d(z^*, Tz^*) + K(\varepsilon)^\alpha \psi(\varepsilon) \end{aligned}$$

for some $K > 0$. It implies that $d(z^*, Tz^*) = 0$, a contradiction. Hence $z^* = Tz^*$.

As a final step, we examine the uniqueness of the found fixed point z^*. Suppose that v^* is another fixed point of T that is distinct from z^*. $Tz^* = z^*$ and $Tv^* = v^*$. By (v) we have

$$\alpha(z^*, Tz^*) \geq 1 \text{ and } \alpha(v, Tv^*) \geq 1.$$

Since $\tfrac{1}{2}d(z^*, Tz^*) \leq d(z^*, v^*)$ the assumption (i) yields that

$$\begin{aligned} d(Tz^*, Tv^*) &\leq (1-\varepsilon) \max\left\{ d(z^*, v^*), d(z, Tz^*), d(v^*, Tv^*), \tfrac{1}{2}[d(z^*, Tv) + d(v^*, Tz^*)] \right\} \\ &\quad + \Lambda(\varepsilon)^\alpha \psi(\varepsilon) \left[1 + 2\|z^*\| + 2\|v^*\|\right]^\beta \\ &< (1-\varepsilon) d(z^*, v^*) + K(\varepsilon)^\alpha \psi(\varepsilon) \end{aligned}$$

for some $K > 0$ that yields that $d(z^*, v^*) = 0$, a contradiction. Hence $z^* = v^*$. □

Example 1. *Let $X = [0, \infty)$ and let $d(x, y) = |x - y|$ for all $x, y \in X$. Let $\Lambda = \tfrac{1}{2}$, $\alpha = 1$, $\beta = 1$ and $\psi(\varepsilon) = \varepsilon^{\tfrac{1}{2}}$ for every $\varepsilon \in [0, 1]$ and a mapping $T : X \to X$ be defined by*

$$Tx = \begin{cases} \tfrac{1}{2}x & \text{if } 0 \leq x \leq 1, \\ 2x & \text{if } x > 1. \end{cases}$$

Also, we define a function $\alpha : X \times X \to [0, \infty)$ in the following way

$$\alpha(x, y) = \begin{cases} 1 & \text{if } 0 \leq x, y \leq 1, \\ 0 & \text{otherwise} \end{cases}$$

Also, we have
$$\frac{1}{2} - 1 + \epsilon \leq \frac{1}{2}(1 - 2 + \frac{\epsilon}{2}) \leq \frac{1}{2}(\epsilon)^{\frac{1}{2}}.$$

Now
$$\frac{1}{2}d(x, Tx) = \frac{1}{2}|x - \frac{x}{2}| \leq d(x, y)$$

implies
$$\begin{aligned}
d(Tx, Ty) &=\leq |Tx - Ty| \\
&= |\tfrac{x}{2} - \tfrac{y}{2}| \\
&= \tfrac{1}{2}|x - y| \\
&\leq \tfrac{1}{2}P(x, y) \\
&= (1 - \epsilon)P(x, y) + (\tfrac{1}{2} - 1 + \epsilon)P(x, y) \\
&\leq (1 - \epsilon)P(x, y) + (\tfrac{1}{2} - 1 + \epsilon)\left[1 + \|x\| + \|y\| + \|Tx\| + \|Ty\|\right] \\
&\leq (1 - \epsilon)P(x, y) + (\tfrac{1}{2}\epsilon\epsilon^{\frac{1}{2}})\left[1 + \|x\| + \|y\| + \|Tx\| + \|Ty\|\right]
\end{aligned}$$

Hence, T satisfies all the conditions of theorem and T has a unique fixed point.

Immediate Consequences

In this subsection, we list a few consequences of our main result. These corollaries also indicate how we can conclude further consequences.

If we let $\alpha(x, Tx) = 1$ for all $x \in X$, we get the following results:

Theorem 4. *Let T be a self-mapping on a metric space (X, d). Suppose that $\beta \in [0, \alpha] \wedge \geq 0$ and $\alpha \geq 1$ are fixed constants. A self-mapping T possesses a unique fixed point if $\frac{1}{2}d(\varkappa, T\varkappa) \leq d(\varkappa, y)$ implies*

$$d(T\varkappa, Ty) \leq P(\varkappa, y)$$

where

$$P(\varkappa, y) = (1 - \varepsilon) \max \left\{ d(\varkappa, y), d(\varkappa, T\varkappa), d(y, Ty), \tfrac{1}{2}[d(\varkappa, Ty) + d(y, T\varkappa)] \right\}$$

$$+ \Lambda(\varepsilon)^\alpha \psi(\varepsilon)\left[1 + \|\varkappa\| + \|y\| + \|T\varkappa\| + \|Ty\|\right]^\beta.$$

for all $\varkappa, y \in X$ and for every $\varepsilon \in [0, 1]$.

Let $(X, \preceq)$ be a partially ordered set and d be a metric on X. We say that $(X, \preceq, d)$ is regular if for every nondecreasing sequence $\{\varkappa_n\} \subset X$ such that $\varkappa_n \to x \in X$ as $n \to \infty$, there exists a subsequence $\{\varkappa_{n(k)}\}$ of $\{\varkappa_n\}$ such that $\varkappa_{n(k)} \preceq x$ for all k.

Theorem 5. *Let $(X, \preceq)$ be a partially ordered set and d be a metric on X such that (X, d) is complete. Let $T: X \to X$ be a nondecreasing mapping with respect to $\preceq$. Suppose that $\beta \in [0, \alpha] \wedge \geq 0$ and $\alpha \geq 1$ are fixed constants such that the self-mapping T satisfies the following condition: $\frac{1}{2}d(\varkappa, T\varkappa) \leq d(\varkappa, y)$ implies*

$$d(T\varkappa, Ty) \leq P(\varkappa, y)$$

where

$$P(\varkappa, y) = (1 - \varepsilon) \max \left\{ d(\varkappa, y), d(\varkappa, T\varkappa), d(y, Ty), \tfrac{1}{2}[d(\varkappa, Ty) + d(y, T\varkappa)] \right\}$$

$$+ \Lambda(\varepsilon)^\alpha \psi(\varepsilon)\left[1 + \|\varkappa\| + \|y\| + \|T\varkappa\| + \|Ty\|\right]^\beta.$$

for all $\varkappa, y \in X$ with $\varkappa \preceq y$ and for every $\varepsilon \in [0, 1]$. Suppose also that the following conditions hold:

(i) there exists $x_0 \in X$ such that $x_0 \preceq Tx_0$;

(ii) $(X, \preceq, d)$ is regular.
(iii) T is nondecreasing with respect to $\preceq$ (that is, $\varkappa, y \in X$, $\varkappa \preceq y \Longrightarrow T\varkappa \preceq Ty$.)

Then T has a fixed point.
Moreover, if for all $\varkappa, y \in X$ there exists $z \in X$ such that $\varkappa, \preceq z$ and $y \preceq z$, we have uniqueness of the fixed point.

Proof. Set $\alpha : X \times X \to [0, \infty)$ in a way that

$$\alpha(x,y) = \begin{cases} 1 \text{ if } \varkappa \preceq y \text{ or } \varkappa \succeq y, \\ 0 \text{ otherwise.} \end{cases}$$

It is apparent that T is an α-Suzuki-Pata contractive mapping, i.e.,

$$\alpha(\varkappa, y)d(T\varkappa, Ty) \leq P(\varkappa, y),$$

for all $\varkappa, y \in X$. By assumption, the inequality $\alpha(\varkappa_0, T\varkappa_0) \geq 1$ is observed. In addition, for all $\varkappa, y \in X$, due to the fact that T is nondecreasing, we find

$$\alpha(x,y) \geq 1 \Longrightarrow x \succeq y \text{ or } x \preceq y \Longrightarrow Tx \succeq Ty \text{ or } Tx \preceq Ty \Longrightarrow \alpha(Tx, Ty) \geq 1.$$

Consequently, we note that T is α-admissible. Now, assume that $(X, \preceq, d)$ is regular. Let $\{\varkappa_n\}$ be a sequence in X such that $\alpha(\varkappa_n, \varkappa_{n+1}) \geq 1$ for all n and $\varkappa_n \to x \in X$ as $n \to \infty$. From the regularity hypothesis, there exists a subsequence $\{\varkappa_{n(k)}\}$ of $\{x_n\}$ such that $\varkappa_{n(k)} \preceq x$ for all k. On account of α we derive that $\alpha(\varkappa_{n(k)}, \varkappa) \geq 1$ for all k. Consequently, the existence and uniqueness of the fixed point is derived by Theorem 3. □

3. Application

In this section, we shall consider an application for our main result. Let $X = C[0,1]$ be the space of all continuous functions defined on interval $[0,1]$ with the metric

$$d(x,y) = \sup_{t \in [0,1]} |x(t) - y(t)|.$$

In what follows we shall use Theorem 5 to show that there is a solution to the following integral equation:

$$x(t) = y(t) + \int_0^1 k(t,s,x(s))ds, t \in [0,1] \qquad (4)$$

Assume that $k(t,s,x)$ is continuous. Let $y \in C[0,1]$.
We consider the following conditions:

(a) $k : [0,1] \times [0,1] \times \mathbb{R} \times \mathbb{R} \to \mathbb{R}$ is continuous;
(b) there exists a continuous function $\gamma : [0, \infty] \times \mathbb{R} \to \mathbb{R}$ such that

$$\sup_{t \in [0,1]} \int_0^1 \gamma(t,s) \leq 1;$$

(c) there exists $\varepsilon \in [0,1]$ such that

$$\left|\frac{1}{2}|x(s) - y(s)| - \int_0^1 k(t,s,x(s))ds\right| \leq |x(s) - y(s)|$$

implies

$$|k(t,s,x(s)) - k(t,s,y(s))| \leq (1-\varepsilon)|x(s) - y(s)|,$$

for all $x, y \in X$;

(d) there exists $x_0 \in C([0,1])$ such that for all $t \in [0,1]$, we have

$$\zeta(x_0(t), \int_0^1 k(t,s,x(s))ds) \geq 0,$$

where $\zeta : X \times X \to [0, \infty)$;

(e) For all $t \in [0,1]$, $x, y \in C[0,1]$,

$$\zeta(x(t), y(t)) \geq 0 \Rightarrow \zeta(\int_0^1 k(t,s,x(s))ds, \int_0^1 k(t,s,y(s))ds) \geq 0;$$

(f) If x_n is a sequence in C[0,1] such that $x_n \to x \in C[0,1]$ and $\zeta(x_n, x_{n+1}) \geq 0$ for all n, then $\zeta(x_n, x) \geq 0$ for all n.

Theorem 6. *Suppose that the conditions (a)–(f) are satisfied. Then, the integral Equation (4) has solution in C[0,1].*

Proof. Since k and the function y are continuous, now define an operator

$$\mathcal{T} : C[0,1] \to C[0,1]$$

write the integral Equation (4) in the form $x = \mathcal{T}x$, where

$$\mathcal{T}x(t) = y(t) + \int_0^1 k(t,s,x(s))ds. \tag{5}$$

It follows that

$$\frac{1}{2}|x(s) - y(s)| - \int_0^1 k(t,s,x(s))ds| \leq (1-\varepsilon)|x(s) - y(s)|$$

implies

$$\begin{aligned}
d(\mathcal{T}x,\mathcal{T}y) &= \sup_{t\in[0,1]} |\mathcal{T}x(t) - \mathcal{T}y(t)| \\
&\leq \sup_{t\in[0,1]} \int_0^1 |k(t,s,x(s)) - k(t,s,y(s))| ds \\
&\leq \sup_{s\in[0,1]} \int_0^1 \gamma(t,s) ds |k(t,s,x) - k(t,s,y)| \\
&\leq (1-\varepsilon)|x(s) - y(s)| \\
&\leq (1-\varepsilon) \max\left\{ d(x,y), d(x,\mathcal{T}x), d(y,\mathcal{T}y), \tfrac{1}{2}[d(x,\mathcal{T}y) + d(y,\mathcal{T}x)]\right\} \\
&\quad + \Lambda(\varepsilon)^\alpha \psi(\varepsilon)\left[1 + \|x\| + \|y\| + \|\mathcal{T}x\| + \|\mathcal{T}y\|\right]^\beta \quad \lambda \geq 0 \quad \alpha \geq 1 \text{ and } \beta \in [0,\alpha].
\end{aligned}$$

Define the function $\alpha : C[0,1] \times C[0,1] \to [0,+\infty)$ by

$$\alpha(x,y) = \begin{cases} 1 & \text{if } \zeta(x(t),y(t)) \geq 0, t \in [0,1], \\ 0 & \text{otherwise}. \end{cases}$$

For all $x,y \in C[0,1]$, we have

Therefore, all the conditions of Theorem 5 are satisfied. Consequently, the mapping $\mathcal{T}$ has a unique fixed point in X, which is a solution of integral equation. □

4. Conclusions

In this paper, we combine and extend significant fixed-point results, namely Suzuki [2], Popescu [1], and Pata [4] by involving the admissible mappings. As in [3] (see also [19]), by proper choice of the auxiliary admissible mapping α and replacing the set $P(x,y)$ with some concrete subset, we can derive several more consequences. Since the techniques are the same in [3], we skip the details and we avoid listing all possible corollaries. Indeed, Theorem 4 and Theorem 5 are the basic examples of this consideration. Notice also that the given example and an integral equation can be improved according to choice of α.

Author Contributions: O.A. analyzed and prepared the manuscript, V.M.L. H.B. analyzed and prepared/edited the manuscript, E.K. analyzed and prepared/edited the manuscript, All authors read and approved the final manuscript.

Funding: This research received no external funding.

Acknowledgments: The authors are grateful to the handling editor and reviewers for their careful reviews and useful comments. The authors would like to extend their sincere appreciation to the Deanship of Scientific Research at King Saud University for funding this group No. RG-1440-025.

Conflicts of Interest: The authors declare no conflict of interest.

References

1. Popescu, O. Two Generalizations of Some Fixed Point Theorems. *Comput. Math. Appl.* **2011**, *62*, 3912–3919. [CrossRef]
2. Suzuki, T. A generalized Banach contraction principle which characterizes metric completeness. *Proc. Am. Math. Soc.* **2008**, *136*, 1861–1869. [CrossRef]
3. Karapinar, E.; Erhan, I.M.; Aksoy, U. Weak ψ-contractions on partially ordered metric spaces and applications to boundary value problems. *Bound. Value Probl.* **2014**, *2014*, 149. [CrossRef]
4. Pata, V. A fixed point theorem in metric spaces. *J. Fixed Point Theory Appl.* **2011**, *10*, 299–305. [CrossRef]
5. Balasubramanian, S. A Pata-type fixed point theorem. *Math. Sci.* **2014**, *8*, 65–69. [CrossRef]
6. Choudhury, B.S.; Metiya, N.; Bandyopadhyay, C.; Maity, P. Fixed points of multivalued mappings satisfying hybrid rational Pata-type inequalities. *J. Anal.* **2018**. [CrossRef]

7. Choudhury, B.S.; Kadelburg, Z.; Metiya, N. Stojan Radenović, S. A Survey of Fixed Point Theorems Under Pata-Type Conditions. *Bull. Malays. Math. Sci. Soc.* **2019**. [CrossRef]
8. Choudhury, B.S.; Metiya, N.; Kundu, S. End point theorems of multivalued operators without continuity satisfying hybrid inequality under two different sets of conditions. *Rendiconti Circolo Matematico Palermo Ser.* **2019**, *68*, 65–81.
9. Geno, K.J.; Khan, M.S.; Choonkil Park, P.; Sungsik, Y. On Generalized Pata Type Contractions. *Mathematics* **2018**, *6*, 25.
10. Eshaghi, M.; Mohseni, S.; Delavar, M.R.; De La Sen, M.; Kim, G.H.; Arian, A. Pata contractions and coupled type fixed points. *Fixed Point Theory Appl.* **2014**, *2014*, 130. [CrossRef]
11. Kadelburg, Z.; Radenovic, S. Fixed point theorems under Pata-type conditions in metric spaces. *J. Egypt. Math. Soc.* **2016**, *24*, 77–82. [CrossRef]
12. Kadelburg, Z.; Radenovic, S. A note on Pata-type cyclic contractions. *Sarajevo J. Math.* **2015**, *11*, 235–245.
13. Kadelburg, Z.; Radenovic, S. Pata-type common fixed point results in b-metric and b-rectangular metric spaces. *J. Nonlinear Sci. Appl.* **2015**, *8*, 944–954. [CrossRef]
14. Kadelburg, Z.; Radenovic, S. Fixed point and tripled fixed point theprems under Pata-type conditions in ordered metric spaces. *Int. J. Anal. Appl.* **2014**, *6*, 113–122.
15. Kolagar, S.M.; Ramezani, M.; Eshaghi, M. Pata type fixed point theorems of multivalued operators in ordered metric spaces with applications to hyperbolic differential inclusions. *Proc. Am. Math. Soc.* **2016**, *6*, 21–34.
16. Ramezani, M.; Ramezani, H. A new generalized contraction and its application in dynamic programming. *Cogent Math.* **2018**, *5*, 1559456. [CrossRef]
17. Popescu, O. Some new fixed point theorems for α-Geraghty contractive type maps in metric spaces. *Fixed Point Theory Appl.* **2014**, *2014*, 190. [CrossRef]
18. Radenovic, S.; Kadelburg, Z.; Jandrlic, D.; Jandrlic, A. Some results on weakly contractive maps. *Bull. Iran. Math. Soc.* **2012**, *38*, 625–645.
19. Karapinar, E.; Samet, B. Generalized (alpha-psi) contractive type mappings and related fixed point theorems with applications. *Abstr. Appl. Anal.* **2012**, *2012*, 793486. [CrossRef]

© 2019 by the authors. Licensee MDPI, Basel, Switzerland. This article is an open access article distributed under the terms and conditions of the Creative Commons Attribution (CC BY) license (http://creativecommons.org/licenses/by/4.0/).

Article

Modified Suzuki-Simulation Type Contractive Mapping in Non-Archimedean Quasi Modular Metric Spaces and Application to Graph Theory

Ekber Girgin * and Mahpeyker Öztürk

Department of Mathematics, Sakarya University, Sakarya 54050, Turkey
* Correspondence: ekber.girgin2@ogr.sakarya.edu.tr

Received: 15 July 2019; Accepted: 16 August 2019; Published: 21 August 2019

Abstract: In this paper, we establish generalized Suzuki-simulation-type contractive mapping and prove fixed point theorems on non-Archimedean quasi modular metric spaces. As an application, we acquire graphic-type results.

Keywords: non-Archimedean quasi modular metric space; θ-contraction; Suzuki contraction; simulation contraction

1. Introduction

In the sequel, the letter $\mathbb{R}_+$ will denote the set of all nonnegative real numbers.
Let S be a nonempty set and $V : S \to S$ be given mappings. A point $\jmath \in S$ is said to be:

i. a fixed point of V if and only if $V\jmath = \jmath$;
ii. a common fixed point of V and Z if and only if $V\jmath = Z\jmath = \jmath$.

Kosjasteh et al. [1] defined a new control function as follows.

Definition 1 ([1]). *Let $\zeta : [0, \infty)^2 \to \mathbb{R}$ be a mapping. The mapping ζ is named a simulation function satisfying the following conditions:*

ζ_1. $\zeta(0,0) = 0$,
ζ_2. $\zeta(a,b) < a - b$, *for all* $a, b > 0$,
ζ_3. *if* $\{a_k\}$ *and* $\{b_k\}$ *are sequences in* $\mathbb{R}_+$ *such that* $\lim_{k \to \infty} a_k = \lim_{k \to \infty} b_k = l, l \in \mathbb{R}_+$. *Thus,*

$$\limsup_{k \to \infty} \zeta(a_k, b_k) < 0.$$

Argoubi et al. [2] modified the above and so introduced it as follows.

Definition 2 ([2]). *The mapping ζ is a simulation function providing the following:*

i. $\zeta(a,b) < a - b$, *for all* $a, b > 0$,
ii. *if* $\{a_k\}$ *and* $\{b_k\}$ *are sequences in* $\mathbb{R}_+$ *such that* $\lim_{k \to \infty} a_k = \lim_{k \to \infty} b_k > 0$, *and* $a_k < b_k$, *then* $\limsup_{k \to \infty} \zeta(a_k, b_k) < 0$.

For examples and related results on simulation functions, one may refer to [1–8].
Radenovic and Chandok generalized the simulation function combining the C-class function as follows.

Definition 3 ([4]). *A mapping $G : [0, \infty)^2 \to \mathbb{R}$ is named a C-class function if it is continuous and satisfies the following conditions:*

i. $G(a,b) \leq a$,
ii. $G(a,b) = a$ implies that either $a = 0$ or $b = 0$, for all $a, b \in [0, \infty)$.

Definition 4 ([4]). *A C_G-simulation function is a mapping $\zeta : [0, \infty)^2 \to \mathbb{R}$ satisfying the following conditions:*

i. $\zeta(a,b) < G(a,b)$ for all $a, b > 0$, where $G : [0, \infty)^2 \to \mathbb{R}$ is a C-class function,
ii. if $\{a_k\}$ and $\{b_k\}$ are sequences in $(0, \infty)$ such that $\lim_{k \to \infty} b_k = \lim_{k \to \infty} a_k > 0$, and $b_k < a_k$, then $\limsup_{k \to \infty} \zeta(a_k, b_k) < C_G$.

Definition 5 ([4]). *A mapping $G : [0, \infty)^2 \to \mathbb{R}$ has the property C_G, if there exists a $C_G \geq 0$ such that:*

i. $G(a,b) > C_G$ implies $a > b$,
ii. $G(a,a) \leq C_G$ for all $a \in [0, \infty)$.

Moreover, using C-class function many researchers investigated some new results combining other control functions in different spaces [9].

Suzuki [10] proved the following fixed point theorem using a new contraction, which is known as the Suzuki contraction in literature. Furthermore, many mathematicians generalized this contraction in other spaces.

Theorem 1 ([10]). *Let (S, d) be a compact metric space and $V : S \to S$ be a mapping. Suppose that, for all $j, \ell \in S$ with $j \neq \ell$,*

$$\frac{1}{2} d(j, Vj) < d(j, \ell) \quad \Rightarrow \quad d(Vj, V\ell) < d(j, \ell).$$

Then, V has a unique fixed point in S.

Bindu et al. [11] proved the commonfixed point theorem for Suzuki type mapping in a complete subspace of the partial metric space.

Theorem 2. *Let (S, δ) be a partial metric space and $f, g, V, Z : S \to S$ be mappings satisfying:*

$$\frac{1}{2} \min\{\delta(fj, Vj), \delta(g\ell, Z\ell)\} \leq \ell(fj, g\ell) \quad \Rightarrow \quad \phi(Vj, Z\ell) \leq \alpha(M(j, \ell)) - \beta(M(j, \ell)),$$

for all $j, \ell \in S$, where $\phi, \alpha, \beta : [0, \infty) \to [0, \infty)$ are such that ϕ is an altering distance function, α is continuous, and β is lower-semi continuous $\alpha(0) = \beta(0) = 0$ and $\phi(t) - \alpha(t) + \beta(t) > 0$, for all $t > 0$ and:

$$M(j, \ell) = \max\left\{\delta(fj, g\ell), \delta(fj, V\ell), \delta(g\ell, Z\ell), \frac{\delta(fj, Z\ell) + \delta(g\ell, Vj)}{2}\right\},$$

i. $V(S) \subseteq g(S)$, $Z(S) \subseteq f(S)$;
ii. either $f(S)$ or $g(S)$ is a complete subspace of S;
iii. the pairs (f, V) and (g, Z) are weakly compatible.

Then, f, g, V, Z have a common fixed point.

Jleli and Samet [12] introduced a Σ-contraction and established fixed point results in generalized metric spaces. Jleli and Samet [12] also introduced a class of Θ such that $\Sigma : (0, \infty) \to (1, \infty)$ of all functions, providing the following conditions:

Σ_1. Σ is nondecreasing;
Σ_2. for any sequence $\{a_k\}$ in $(0, \infty)$, $\lim_{k \to \infty} \Sigma(a_k) = 1$ if and only if $\lim_{k \to \infty} a_k = 0$;

Σ_3. there exist $r \in (0,1)$ and $l \in (0,\infty)$ such that $\lim_{k \to 0^+} \frac{\Sigma(k)-1}{k^r} = l$.

Theorem 3 ([12]). *Let (S,d) be a complete generalized metric space and $V : S \to S$ be a mapping. Suppose that there exist $\Sigma \in \Theta$ and $\gamma \in (0,1)$ such that:*

$$d(V\jmath, V\ell) \neq 0 \Rightarrow \Sigma(d(V\jmath, V\ell)) \leq [\Sigma(d(\jmath,\ell))]^\gamma,$$

for all $\jmath, \ell \in S$. Then, V has a unique fixed point.

After that, many authors generalized such a contraction in different spaces [13–17].

Liu et al. [15] modified the class of function Θ exchanging conditions. The class of functions $\tilde{\Theta}$ was defined by the set of $\Sigma : (0,\infty) \to (1,\infty)$ satisfying the following conditions:

$\tilde{\Sigma}_1$. Σ is non-decreasing and continuous,
$\tilde{\Sigma}_2$. $\inf_{k \in (0,\infty)} \Sigma(k) = 1$.

Lemma 1 ([15]). *Let $\Sigma : (0,\infty) \to (1,\infty)$ be a non-decreasing and continuous function with $\inf_{k \in (0,\infty)} \Sigma(k) = 1$ and $\{a_k\}$ be a sequence in $(0,\infty)$. Then, the following condition holds:*

$$\lim_{k \to \infty} \Sigma(a_k) = 1 \Leftrightarrow \lim_{k \to \infty} a_k = 0.$$

Zheng et al. [18] denoted new set functions Φ satisfying the following conditions:

Φ_1. $\varphi : [1,\infty) \to [1,\infty)$ is nondecreasing,
Φ_2. for each $k > 0$, $\lim_{n \to \infty} \varphi^n(k) = 1$,
Φ_3. φ is continuous on $[1,\infty)$.

Lemma 2 ([18]). *If $\varphi \in \Phi$, then $\varphi(1) = 1$ and $\varphi(t) < t$ for each $t > 1$.*

Definition 6 ([18]). *Let (S,d) be a metric space and $V : S \to S$ be a mapping. V is said to be a $\Sigma - \varphi$-contraction if there exist $\Sigma \in \Theta$ and $\varphi \in \Phi$ such that for any $\jmath, \ell \in S$,*

$$\Sigma(d(V\jmath, V\ell)) \leq \varphi(\Sigma(N(\jmath,\ell))),$$

where:

$$N(\jmath,\ell) = \max\{d(\jmath,\ell), d(\jmath, V\ell), d(\jmath, V\ell)\}.$$

Theorem 4 ([18]). *Let (S,d) be a complete metric space and $V : S \to S$ be a $\Sigma - \varphi$-contraction. Then, V has a unique fixed point.*

Motivated by the above, we will establish a generalized Suzuki-simulation-type contractive mapping and obtain fixed point results.

2. Quasi Modular Metric Space

Girgin and Öztürk [19] introduced a new space, which is named a quasi modular metric space. Furthermore, they gave some topological properties. Moreover, defining non-Archimedean quasi modular metric space, they proved some fixed point theorems and obtained some applications.

Definition 7 ([19]). *A function $Q : (0,\infty) \times S \times S \to [0,\infty]$ is called a quasi modular metric on S if the following hold:*

q_1. $\xi = \eta$ if and only if $Q_m(\xi,\eta) = 0$ for all $m > 0$;

q_2. $Q_{m+n}(\xi, \eta) \leq Q_m(\xi, \nu) + Q_n(\nu, \eta)$ for all $m, n > 0$ and $\xi, \eta, \nu \in S$.

Then, S_Q is a quasi modular metric space. If in the above definition, we utilize the condition:

$q_{1'}$. $Q_m(\xi, \xi) = 0$ for all $m > 0$ and $\xi \in S$,

instead of (q_1), then Q is said to be a quasi pseudo modular metric on S. A quasi modular metric Q on S is called a regular if the following weaker version of (q_1) is satisfied:

q_3. $\xi = \eta$ if and only if $Q_m(\xi, \eta) = 0$ for some $m > 0$.

Again, Q is called a convex if for $m, n > 0$ and $\xi, \eta, \nu \in S$, the inequality holds:

q_4. $Q_{m+n}(\xi, \eta) \leq \frac{m}{m+n} Q_m(\xi, \nu) + \frac{n}{m+n} Q_n(\nu, \eta)$.

Definition 8 ([19]). *In Definition 7, if we replace (q_2) by:*

q_5. $Q_{\max\{m,n\}}(\xi, \eta) \leq Q_m(\xi, \nu) + Q_n(\nu, \eta)$

for all $m, n > 0$ and $\xi, \eta, \nu \in S$, then S_Q is called a non-Archimedean quasi modular metric space.

Note that the function $Q_{\max\{m,n\}}$ is more general than the function $Q_{m+n}(\xi, \eta)$, so every non-Archimedean quasi modular metric space is a quasi modular metric space.

Example 1 ([19]). *Let $S = [0, \infty)$ and Q be defined by:*

$$Q_m(\xi, \eta) = \begin{cases} \frac{\xi - \eta}{m} & \text{if } \xi \geq \eta \\ 1 & \text{if } \xi < \eta. \end{cases}$$

Then, S_Q is a quasi modular metric space with $m = \frac{1}{3}$ and $n = \frac{2}{3}$, but is not modular metric space since $Q_m(0,1) = 1$ and $Q_m(1,0) = \frac{1}{3}$.

Remark 1 ([19]). *From the above definitions we deduce that:*

i. For a quasi modular metric Q on S, the conjugate quasi modular metric Q^{-1} on S of Q is defined by $Q_m^{-1}(\xi, \eta) = Q_m(\eta, \xi)$.
ii. If Q is a T_0-quasi pseudo modular metric on S, then the function Q^E defined by $Q^E = Q^{-1} \vee Q$, that is $Q_m^E(\xi, \eta) = \max\{Q_m(\xi, \eta), Q_m(\eta, \xi)\}$, defines a modular metric space.

Now, we discuss some topological properties of quasi modular metric spaces.

Definition 9 ([19]). *A sequence $\{\xi_p\}$ in S_Q converges to ξ and is called:*

a. Q-convergent or left convergent if $\xi_p \to \xi \Leftrightarrow Q_m(\xi, \xi_p) \to 0$.
b. Q^{-1}-convergent or right convergent if $\xi_p \to \xi \Leftrightarrow Q_m(\xi_p, \xi) \to 0$.
c. Q^E-convergent if $Q_m(\xi, \xi_p) \to 0$ and $Q_m(\xi_p, \xi) \to 0$.

Definition 10 ([19]). *A sequence $\{\xi_p\}$ in a quasi modular metric space S_Q is called:*

d. left (right) Q-K-Cauchy if for every $\varepsilon > 0$, there exists $p_\varepsilon \in N$ such that $Q_m(\xi_r, \xi_p) < \varepsilon$ for all $p, r \in N$ with $p_\varepsilon \leq r \leq p$ ($p_\varepsilon \leq p \leq r$) and for all $m > 0$.
e. Q^E-Cauchy if for every $\varepsilon > 0$, there exists $p_\varepsilon \in N$ such that $Q_m(\xi_p, \xi_r) < \varepsilon$ for all $p, r \in N$ with $p, r \geq p_\varepsilon$.

Remark 2 ([19]). *From the above definitions, we deduce that:*

i. a sequence is left Q-K-Cauchy with respect to Q if and only if it is right Q-K-Cauchy with respect to Q^{-1};

ii. a sequence is Q^E-Cauchy if and only if it is left and right Q-K-Cauchy.

Definition 11 ([19]). *A quasi modular metric space S_Q is called:*

i. *left Q-K-complete if every left Q-K-Cauchy is Q-convergent.*
ii. *Q-Smyth-complete if every left Q-K-Cauchy sequence is Q^E-convergent.*

3. Common Fixed Point Results

In the sequel, Q is regular and convex and T_Z denotes the family of all C_G-simulation functions $\zeta : [0, \infty)^2 \to \mathbb{R}$.

Definition 12. *Let S_Q be a non-Archimedean quasi modular metric space and $V : S_Q \to S_Q$ be a mapping. We say that V is a generalized Suzuki-simulation-type contractive mapping if there exist $\Sigma \in \tilde{\Theta}$, $\varphi \in \Phi$ and $\zeta \in T_Z$ such that:*

$$\tfrac{1}{2} Q_m(\xi, V\xi) \leq Q_m(\xi, \eta) \quad \text{implies} \tag{1}$$

$$\zeta\left(\Sigma\left(Q_m(V\xi, V\eta)\right), \varphi\left(\Sigma\left(P(\xi, \eta)\right)\right)\right) \geq C_G$$

where:

$$P(\xi, \eta) = \max\{Q_m(\xi, \eta), Q_m(\xi, V\xi), Q_m(\eta, V\eta)\}$$

for all $\xi, \eta \in S_Q$.

Theorem 5. *Let S_Q be a Q-Smyth-complete non-Archimedean quasi modular metric space and V be the generalized Suzuki-simulation-type contractive mapping. Then, V has a unique fixed point.*

Proof. Define a sequence $\{\xi_k\}$ in S_Q by:

$$\xi_{k+1} = V\xi_k, \tag{2}$$

for all $k \in \mathbb{N}$. If there exists an k_0 such that $\xi_{k_0} = \xi_{k_0+1}$, then $z = \xi_{k_0}$ becomes a fixed point of V. Consequently, we shall assume that $\xi_k \neq \xi_{k+1}$ for all $k \in \mathbb{N}$. Therefore, we have:

$$Q_m(\xi_k, \xi_{k+1}) > 0, \quad \text{for all } n = 0, 1, 2\ldots. \tag{3}$$

Hence, we have:

$$\frac{1}{2} Q_m(\xi_k, V\xi_k) < Q_m(\xi_k, V\xi_k) = Q_m(\xi_k, \xi_{k+1}) \quad \text{implies,}$$

$$C_G \leq \zeta\left(\Sigma\left(Q_m(V\xi_k, V\xi_{k+1})\right), \varphi\left(\Sigma\left(P(\xi_k, \xi_{k+1})\right)\right)\right)$$

$$= \zeta\left(\Sigma\left(Q_m(\xi_{k+1}, \xi_{k+2})\right), \varphi\left(\Sigma\left(P(\xi_k, \xi_{k+1})\right)\right)\right) \tag{4}$$

$$< G\left(\varphi\left(\Sigma\left(P(\xi_k, \xi_{k+1})\right)\right), \Sigma\left(Q_m(\xi_{k+1}, \xi_{k+2})\right)\right),$$

by Definition 5, we get that:

$$\Sigma\left(Q_m(\xi_{k+1}, \xi_{k+2})\right) < \varphi\left(\Sigma\left(P(\xi_k, \xi_{k+1})\right)\right), \tag{5}$$

where:
$$P(\xi_k, \xi_{k+1}) = \max\{Q_m(\xi_k, \xi_{k+1}), Q_m(\xi_k, V\xi_k), Q_m(\xi_{k+1}, V\xi_{k+1})\}$$
$$= \max\{Q_m(\xi_k, \xi_{k+1}), Q_m(\xi_k, \xi_{k+1}), Q_m(\xi_{k+1}, \xi_{k+2})\} \quad (6)$$
$$= \max\{Q_m(\xi_k, \xi_{k+1}), Q_m(\xi_{k+1}, \xi_{k+2})\}.$$

If:
$$\max\{Q_m(\xi_k, \xi_{k+1}), Q_m(\xi_{k+1}, \xi_{k+2})\} = Q_m(\xi_{k+1}, \xi_{k+2})$$

for some $k \in \mathbb{N}$, then it follows from (5) and Lemma 2 that we get:

$$\Sigma(Q_m(\xi_{k+1}, \xi_{k+2})) < \varphi(\Sigma(Q_m(\xi_{k+1}, \xi_{k+2}))) < \Sigma(Q_m(\xi_{k+1}, \xi_{k+2}))$$

which is a contradiction. Therefore, we have:

$$P(\xi_k, \xi_{k+1}) = Q_m(\xi_k, \xi_{k+1})$$

for each $k \in \mathbb{N}$. Also, by (5), we have

$$\Sigma(Q_m(\xi_{k+1}, \xi_{k+2})) < \varphi(\Sigma(Q_m(\xi_k, \xi_{k+1}))).$$

Repeating this step, we conclude that:

$$\Sigma(Q_m(\xi_{k+1}, \xi_{k+2})) < \varphi(\Sigma(Q_m(\xi_k, \xi_{k+1})))$$
$$< \varphi^2(\Sigma(Q_m(\xi_{k-1}, \xi_k)))$$
$$\vdots$$
$$< \varphi^k(\Sigma(Q_m(\xi_1, \xi_2))),$$

for all $k \in \mathbb{N}$. Taking the limit $k \to \infty$ above, by the definition of φ and property Θ_2, we have:

$$\lim_{k \to \infty} \varphi^k(Q_m(\xi_1, \xi_2)) = 1. \quad (7)$$

Thus, from Lemma 1, it follows that:

$$\lim_{k \to \infty} Q_m(\xi_{k+1}, \xi_{k+2}) = 0, \quad (8)$$

for all $k \in \mathbb{N}$. Now, we show that $\{\xi_k\}$ is a left Q-K-Cauchy sequence. Assume the contrary. There exists $\varepsilon > 0$ such that we can find two subsequences $\{t_k\}$ and $\{s_k\}$ of positive integers satisfying the following inequalities:

$$Q_m(\xi_{t_k}, \xi_{s_k}) \geq \varepsilon, \text{ and } Q_m(\xi_{t_k-1}, \xi_{s_k}) < \varepsilon. \quad (9)$$

From (9) and (q_5), it follows that:

$$\varepsilon \leq Q_m(\xi_{t_k}, \xi_{s_k}) = Q_{\max\{m,m\}}(\xi_{t_k}, \xi_{s_k})$$
$$\leq Q_m(\xi_{t_k}, \xi_{t_k-1}) + Q_m(\xi_{t_k-1}, \xi_{s_k}) \quad (10)$$
$$< \varepsilon + Q_m(\xi_{t_k}, \xi_{t_k-1}).$$

On taking the limit as $k \to \infty$ in the above relation, we obtain that:

$$\lim_{k \to \infty} Q_m(\xi_{t_k}, \xi_{s_k}) = \varepsilon. \quad (11)$$

Also, from (9) and (q_5), it follows that:

$$\begin{aligned}Q_m\left(\xi_{t_k+1},\zeta_{s_k+1}\right) &= Q_{\max\{m,m\}}\left(\xi_{t_k+1},\zeta_{s_k+1}\right) \\ &\leq Q_m\left(\xi_{t_k+1},\xi_{t_k}\right) + Q_m\left(\xi_{t_k},\zeta_{s_k+1}\right) \\ &= Q_m\left(\xi_{t_k+1},\xi_{t_k}\right) + Q_{\max\{m,m\}}\left(\xi_{t_k},\zeta_{s_k+1}\right) \\ &\leq Q_m\left(\xi_{t_k},\xi_{t_k-1}\right) + Q_m\left(\xi_{t_k-1},\zeta_{s_k+1}\right) + Q_m\left(\xi_{t_k+1},\xi_{t_k}\right) \\ &= Q_m\left(\xi_{t_k},\xi_{t_k-1}\right) + Q_m\left(\xi_{t_k+1},\xi_{t_k}\right) + Q_{\max\{m,m\}}\left(\xi_{t_k-1},\zeta_{s_k+1}\right) \\ &\leq Q_m\left(\xi_{t_k-1},\zeta_{s_k}\right) + Q_m\left(\zeta_{s_k},\zeta_{s_k+1}\right) \\ &\quad + Q_m\left(\xi_{t_k},\xi_{t_k-1}\right) + Q_m\left(\xi_{t_k+1},\xi_{t_k}\right) \\ &< \varepsilon + Q_m\left(\zeta_{s_k},\zeta_{s_k+1}\right) + Q_m\left(\xi_{t_k},\xi_{t_k-1}\right) \\ &\quad + Q_m\left(\xi_{t_k+1},\xi_{t_k}\right).\end{aligned}\qquad(12)$$

Next, we claim that:
$$\frac{1}{2}Q_m\left(\xi_{t_k},V\xi_{t_k}\right) \leq Q_m\left(\xi_{t_k},\zeta_{s_k}\right).$$

If:
$$\begin{aligned}\tfrac{1}{2}Q_m\left(\xi_{t_k},V\xi_{t_k}\right) &> Q_m\left(\xi_{t_k},\zeta_{s_k}\right) \\ &= \tfrac{1}{2}Q_m\left(\xi_{t_k},\xi_{t_k+1}\right) > Q_m\left(\xi_{t_k},\zeta_{s_k}\right),\end{aligned}\qquad(13)$$

then letting $k \to \infty$ in (13), from (11) and (8), we have that $0 > \varepsilon$ is a contradiction. Hence,

$$\frac{1}{2}Q_m\left(\xi_{t_k},V\xi_{t_k}\right) \leq Q_m\left(\xi_{t_k},\zeta_{s_k}\right).$$

From the generalized Suzuki-simulation-type contractive mapping, we get:

$$\begin{aligned}C_G &\leq \zeta\left(\Sigma\left(Q_m\left(V\xi_{t_k},V\zeta_{s_k}\right)\right),\varphi\left(\Sigma\left(P\left(\xi_{t_k},\zeta_{s_k}\right)\right)\right)\right) \\ &= \zeta\left(\Sigma\left(Q_m\left(\xi_{t_k+1},\zeta_{s_k+1}\right)\right),\varphi\left(\Sigma\left(P\left(\xi_{t_k},\zeta_{s_k}\right)\right)\right)\right),\end{aligned}\qquad(14)$$

where:

$$\begin{aligned}P\left(\xi_{t_k},\zeta_{s_k}\right) &= \max\left\{Q_m\left(\xi_{t_k},\zeta_{s_k}\right),Q_m\left(\xi_{t_k},V\xi_{t_k}\right),Q_m\left(\zeta_{s_k},V\zeta_{s_k}\right)\right\} \\ &= \max\left\{Q_m\left(\xi_{t_k},\zeta_{s_k}\right),Q_m\left(\xi_{t_k},\xi_{t_k+1}\right),Q_m\left(\zeta_{s_k},\zeta_{s_k+1}\right)\right\}.\end{aligned}\qquad(15)$$

Taking the limit $k \to \infty$ using (8), (11), and (12) in (14) and (15), we obtain:

$$C_G \leq \zeta\left(\Sigma\left(\varepsilon\right),\varphi\left(\Sigma\left(\varepsilon\right)\right)\right) < G\left(\varphi\left(\Sigma\left(\varepsilon\right)\right),\Sigma\left(\varepsilon\right)\right).$$

From Definition 5, we get:
$$\Sigma\left(\varepsilon\right) < \varphi\left(\Sigma\left(\varepsilon\right)\right) < \Sigma\left(\varepsilon\right).$$

It follows that $\Sigma\left(\varepsilon\right) < \Sigma\left(\varepsilon\right)$, a contradiction. Hence, $\{\xi_k\}$ is a left Q-K-Cauchy sequence. As S_Q is a Q-Smyth-complete non-Archimedean quasi modular metric space, there exists $u \in S_Q$ such that:

$$\lim_{k\to\infty} Q_m{}^E\left(\xi_k,u\right) = 0.$$

Thus, we have:
$$\lim_{k\to\infty} Q_m\left(\xi_k,u\right) = 0 \quad \text{and} \quad \lim_{k\to\infty} Q_m\left(u,\xi_k\right) = 0.$$

Now, we show that u is a fixed point of V. Assume that $Q_m\left(Vu,u\right) > 0$. We claim that for each $k \geq 0$, the following holds:

$$\frac{1}{2}Q_m\left(\xi_k,V\xi_k\right) \leq Q_m\left(\xi_k,u\right).$$

On the contrary, suppose that:

$$\frac{1}{2} Q_m(\xi_k, V\xi_k) > Q_m(\xi_k, u) = \frac{1}{2} Q_m(\xi_k, \xi_{k+1}) > Q_m(\xi_k, u). \quad (16)$$

Taking the limit as $k \to \infty$ in (16), we obtain $0 > 0$, a contradiction. Thus, the claim is true, and so, from the generalized Suzuki-simulation-type contractive mapping, we get:

$$C_G \leq \zeta\left(\Sigma\left(Q_m(V\xi_k, Vu)\right), \varphi\left(\Sigma\left(P(\xi_k, u)\right)\right)\right)$$
$$= \zeta\left(\Sigma\left(Q_m(\xi_{k+1}, Vu)\right), \varphi\left(\Sigma\left(P(\xi_k, u)\right)\right)\right) \quad (17)$$
$$< G\left(\varphi\left(\Sigma\left(P(\xi_k, u)\right)\right), \Sigma\left(Q_m(\xi_{k+1}, Vu)\right)\right).$$

By Definition 5,

$$\Sigma\left(Q_m(\xi_{k+1}, Vu)\right) < \varphi\left(\Sigma\left(P(\xi_k, u)\right)\right), \quad (18)$$

where:

$$P(\xi_k, u) = \max\left\{Q_m(\xi_k, u), Q_m(\xi_k, V\xi_k), Q_m(u, Vu)\right\} \quad (19)$$
$$= \max\left\{Q_m(\xi_k, u), Q_m(\xi_k, \xi_{k+1}), Q_m(u, Vu)\right\}.$$

Letting $k \to \infty$ in (17)–(19), we have:

$$\Sigma\left(Q_m(u, Vu)\right) < \varphi\left(\Sigma\left(Q_m(u, Vu)\right)\right) < \Sigma\left(Q_m(u, Vu)\right).$$

That is, $\Sigma\left(Q_m(u, Vu)\right) < \Sigma\left(Q_m(u, Vu)\right)$, a contradiction. Thus, u is a fixed point of V. Suppose that there is an another fixed point u^* of V such that $Vu^* = u^*$ and $u \neq u^*$. Then, $Q_m(Vu, Vu^*) = Q_m(u, u^*) > 0$, and:

$$0 = \frac{1}{2} Q_m(u, Vu) \leq Q_m(u, u^*).$$

By the generalized Suzuki-simulation-type contractive mapping, we have:

$$C_G \leq \zeta\left(\Sigma\left(Q_m(Vu, Vu^*)\right), \varphi\left(\Sigma\left(P(u, u^*)\right)\right)\right)$$
$$= \zeta\left(\Sigma\left(Q_m(u, u^*)\right), \varphi\left(\Sigma\left(P(u, u^*)\right)\right)\right) \quad (20)$$
$$< G\left(\varphi\left(\Sigma\left(P(u, u^*)\right)\right), \Sigma\left(Q_m(u, u^*)\right)\right).$$

From the property of G,

$$\Sigma\left(Q_m(u, u^*)\right) < \varphi\left(\Sigma\left(P(u, u^*)\right)\right), \quad (21)$$

where:

$$P(u, u^*) = \max\left\{Q_m(u, u^*), Q_m(u, Vu), Q_m(u^*, Vu^*)\right\} = Q_m(u, u^*). \quad (22)$$

From (20)–(22), we attain the following ordering:

$$\Sigma\left(Q_m(u, u^*)\right) < \varphi\left(\Sigma\left(Q_m(u, u^*)\right)\right) < \Sigma\left(Q_m(u, u^*)\right),$$

which is a contradiction. Hence, u is a unique fixed point of V. □

Now, we give some corollaries that are directly acquired from our main results.

Corollary 1. Let S_Q be a Q-Smyth-complete non-Archimedean quasi modular metric space and $V : S_Q \to S_Q$ be a mapping. If there exists $\Sigma \in \tilde{\Theta}$, $\varphi \in \Phi$, and $\zeta \in T_Z$ such that:

$$\frac{1}{2} Q_m(\jmath, V\jmath) \leq Q_m(\jmath, \ell) \quad \text{implies,}$$

$$\zeta\left(\Sigma\left(Q_m(V\jmath, V\ell)\right), \varphi\left(\Sigma\left(Q_m(\jmath, \ell)\right)\right)\right) \geq C_G,$$

for all $\jmath, \ell \in S_Q$, then V has a unique fixed point.

Corollary 2. Let S_Q be a Q-Smyth-complete non-Archimedean quasi modular metric space and $V : S_Q \to S_Q$ be a mapping. If there exists $\Sigma \in \tilde{\Theta}$, $\varphi \in \Phi$, and $\zeta \in T_Z$ such that:

$$\zeta\left(\Sigma\left(Q_m(V\jmath, V\ell)\right), \varphi\left(\Sigma\left(P(\jmath, \ell)\right)\right)\right) \geq C_G$$

where:

$$P(\jmath, \ell) = \max\{Q_m(\jmath, \ell), Q_m(\jmath, V\jmath), Q_m(\ell, V\ell)\},$$

for all $\jmath, \ell \in S_Q$, then V has a unique fixed point.

Corollary 3. Let S_Q be a Q-Smyth-complete non-Archimedean quasi modular metric space and $V : S_Q \to S_Q$ be a mapping. If there exists $\Sigma \in \tilde{\Theta}$ and $\varphi \in \Phi$ such that:

$$\frac{1}{2} Q_m(\jmath, V\jmath) \leq Q_m(\jmath, \ell) \quad \text{implies,}$$

$$\Sigma(Q_m(V\jmath, V\ell)) \leq \varphi(\Sigma(P(\jmath, \ell)))$$

where:

$$P(\jmath, \ell) = \max\{Q_m(\jmath, \ell), Q_m(\jmath, V\jmath), Q_m(\ell, V\ell)\},$$

for all $\jmath, \ell \in S_Q$, then V has a unique fixed point.

Corollary 4. Let S_Q be a Q-Smyth-complete non-Archimedean quasi modular metric space and $V : S_Q \to S_Q$ be a mapping. If there exists $\Sigma \in \tilde{\Theta}$ and $\varphi \in \Phi$ such that:

$$\Sigma(Q_m(V\jmath, V\ell)) \leq \varphi(\Sigma(P(\jmath, \ell)))$$

where:

$$P(\jmath, \ell) = \max\{Q_m(\jmath, \ell), Q_m(\jmath, V\jmath), Q_m(\ell, V\ell)\},$$

for all $\jmath, \ell \in S_Q$, then V has a unique fixed point.

Corollary 5. Let S_Q be a Q-Smyth-complete non-Archimedean quasi modular metric space and $V : S_Q \to S_Q$ be a mapping. If there exists $\Sigma \in \tilde{\Theta}$ and $\varphi \in \Phi$ such that:

$$\Sigma(Q_m(V\jmath, V\ell)) \leq \varphi(\Sigma(Q_m(\jmath, \ell))),$$

for all $\jmath, \ell \in S_Q$, then V has a unique fixed point.

4. Application to a Graph Structure

Let S_Q be a non-Archimedean quasi modular metric space and $\Delta = \{(\jmath,\jmath) : \jmath \in S_Q\}$ denote the diagonal of $S_Q \times S_Q$. Let H be a directed graph such that the set $C(H)$ of its vertices coincides with S_Q and $B(H)$ is the set of edges of the graph such that $\Delta \subseteq B(H)$. H is determined with the pair $(C(H), B(H))$.

If $ȷ$ and ℓ are vertices of H, then a path in H from $ȷ$ to ℓ of length $p \in \mathbb{N}$ is a finite sequence $\{ȷ_p\}$ of vertices such that $ȷ = ȷ_0, ..., ȷ_p = \eta$ and $(ȷ_{i-1}, ȷ_i) \in B(H)$ for $i \in \{1, 2, ..., p\}$.

Recall that H is connected if there is a path between any two vertices, and it is weakly connected if $\tilde{H}$ is connected, where $\tilde{H}$ defines the undirected graph obtained from H by ignoring the direction of edges. Define by H^{-1} the graph obtained from H by reversing the direction of edges. Thus,

$$B\left(H^{-1}\right) = \{(ȷ, \ell) \in S_Q \times S_Q : (\ell, ȷ) \in B(H)\}.$$

It is more convenient to treat $\tilde{H}$ as a directed graph for which the set of its edges is symmetric, under this convention; we have that:

$$B(\tilde{H}) = B(H) \cup B(H^{-1}).$$

Let $H_ȷ$ be the component of H consisting of all the edges and vertices that are contained in some way in H starting at $ȷ$. We denote the relation (R) in the following way:

We have $ȷ(R)\ell$ if and only if, there is a path in H from $ȷ$ to ℓ, for $ȷ, \ell \in C(H)$.

If H is such that $B(H)$ is symmetric, then for $ȷ \in C(H)$, the equivalence class $[ȷ]_G$ in $V(G)$ described by the relation (R) is $C(H_ȷ)$.

Let S_Q be a non-Archimedean quasi modular metric space endowed with a graph H and $\hbar : S_Q \to S_Q$. We set:

$$S_\hbar = \{ȷ \in S_Q : (ȷ, \hbar ȷ) \in B(H)\}.$$

Definition 13 ([20]). *(S, d) is a metric space, and $\hbar : S \to S$ is a mapping. Then, $\hbar$ is called a Banach H-contraction if the following hold:*

B_1. *$\hbar$ preserves edges of H, i.e., for all $ȷ, \ell \in S$,*

$$(ȷ, \ell) \in B(H) \quad \Rightarrow \quad (\hbar ȷ, \hbar \ell) \in B(H),$$

B_2. *there exists $\delta \in (0, 1)$ such that:*

$$d(\hbar ȷ, \hbar \ell) \leq \delta d(ȷ, \ell)$$

for all $(ȷ, \ell) \in B(H)$.

After that, many fixed point researchers investigated fixed point results improving the Jachymski fixed point theorems in [17,21–23].

Now, motivated by [24–26], we generate a new contraction and obtain fixed point results using a graph structure.

Definition 14. *Let S_Q be a non-Archimedean quasi modular metric space and $\hbar : S_Q \to S_Q$ be a mapping. Then, we say that $\hbar$ is a generalized Suzuki-simulation-H-type contractive mapping if the following conditions hold:*

H_1. *$\hbar$ preserves edges of G;*
H_2. *there exists $\Sigma \in \tilde{\Theta}$, $\varphi \in \Phi$ and $\zeta \in T_Z$ such that:*

$$\tfrac{1}{2} Q_m(ȷ, \hbar ȷ) \leq Q_m(ȷ, \ell) \quad \text{implies,}$$

$$\zeta\left(\Sigma\left(Q_m(\hbar ȷ, \hbar \ell)\right), \varphi\left(\Sigma\left(P(ȷ, \ell)\right)\right)\right) \geq C_G, \tag{23}$$

where

$$P(ȷ, \ell) = \max\{Q_m(ȷ, \ell), Q_m(ȷ, \hbar ȷ), Q_m(\ell, \hbar \ell)\}$$

for all $ȷ, \ell \in B(H)$ and for all $m > 0$.

Remark 3. Let S_Q be a non-Archimedean quasi modular metric space with a graph H and $\hbar : S_Q \to S_Q$ be a generalized Suzuki-simulation-H-type contractive mapping. If there exists $J_0 \in S_Q$ such that $\hbar J_0 \in [J_0]_{\tilde{H}}$, then:

R_1. $\hbar$ is both a generalized Suzuki-simulation-H^{-1}-type contractive mapping and a generalized Suzuki-Simulation-$\tilde{H}$-type contractive mapping.

R_2. $[J_0]_{\tilde{H}}$ is $\hbar$-invariant, and $\hbar \big|_{[J_0]_{\tilde{H}}}$ is a generalized Suzuki-simulation-$\tilde{H}_{J_0}$-type contractive mapping.

Theorem 6. Let S_Q be a Q-Smyth-complete non-Archimedean quasi modular metric space with a graph H and $\hbar : S_Q \to S_Q$ be a mapping.

i. there exists $J_0 \in S_\hbar$;
ii. $\hbar$ is the generalized Suzuki-simulation-$\tilde{H}$-type contractive mapping;
iii. H is weakly connected;
iv. if $\{J_k\}$ is a sequence in S_Q such that $\lim_{k\to\infty} Q_m^E(J_k, u) = 0$ and $(J_k, J_{k+1}) \in B(H)$, then there exists a subsequence $\{J_{k_s}\}$ of $\{J_k\}$ such that $(J_{k_s}, u) \in B(H)$.

Then, $\hbar$ has a unique fixed point.

Proof. Define a sequence $\{J_k\}$ in S_Q by:

$$J_{k+1} = \hbar J_k, \tag{24}$$

for all $k \in \mathbb{N}$. Let J_0 be a given point in $S_\hbar$; thus, $(J_0, \hbar J_0) = (J_0, J_1) \in B(H)$. Because $\hbar$ preserves the edges of H,

$$(J_0, J_1) \in B(H) \quad \Rightarrow \quad (\hbar J_0, \hbar J_1) \in B(H).$$

Continuing this way, we get:

$$(J_k, J_{k+1}) \in B(H).$$

Then from Theorem 5, we get that $\{J_k\}$ is a left Q-K-Cauchy sequence in S_Q. By the Q-Smyth-completeness of S_Q, there exists $u \in S_Q$ such that:

$$\lim_{k\to\infty} Q_m^E(J_k, u) = 0. \tag{25}$$

Thus, we have:

$$\lim_{k\to\infty} Q_m(J_k, u) = 0 \text{ and } \lim_{k\to\infty} Q_m(u, J_k) = 0. \tag{26}$$

Now, we show that u is a fixed point of $\hbar$. Using (iv), we get $(J_{k_s}, u) \in B(H)$. We claim that:

$$\frac{1}{2} Q_m(J_{k_s}, \hbar J_{k_s}) \leq Q_m(J_{k_s}, u). \tag{27}$$

If

$$\frac{1}{2} Q_m(J_{k_s}, \hbar J_{k_s}) > Q_m(J_{k_s}, u) = \frac{1}{2} Q_m(J_{k_s}, J_{k_s+1}) > Q_m(J_{k_s}, u) \tag{28}$$

and taking the limit $s \to \infty$ in (28), we get $0 > 0$, a contradiction. Hence, the claim is true. Since $\hbar$ is a generalized Suzuki-simulation-$\tilde{H}$-type contractive mapping, we have:

$$C_G \leq \zeta\left(\Sigma\left(Q_m(\hbar J_{k_s}, \hbar u)\right), \varphi\left(\Sigma\left(P(J_{k_s}, u)\right)\right)\right)$$

$$\leq \zeta\left(\Sigma\left(Q_m(\hbar J_{k_s}, \hbar u)\right), \varphi\left(\Sigma\left(P(J_{k_s}, u)\right)\right)\right) \tag{29}$$

$$\leq G\left(\varphi\left(\Sigma\left(P(J_{k_s}, u)\right)\right), \Sigma\left(Q_m(h J_{k_s}, hu)\right)\right),$$

from Definition 5, we get:
$$\Sigma\left(Q_m\left(\hbar j_{k_s},\hbar u\right)\right),\varphi\left(\Sigma\left(P\left(j_{k_s},u\right)\right)\right), \quad (30)$$

where:
$$P\left(j_{k_s},u\right)=\max\left\{Q_m\left(j_{k_s},u\right),Q_m\left(j_{k_s},\hbar j_{k_s}\right),Q_m\left(u,\hbar u\right)\right\}$$
$$=\max\left\{Q_m\left(j_{k_s},u\right),Q_m\left(j_{k_s},j_{k_s+1}\right),Q_m\left(u,\hbar u\right)\right\}. \quad (31)$$

Taking the limit as $s\to\infty$ in (29)–(31), we get:
$$\Sigma\left(Q_m\left(u,\hbar u\right)\right)<\varphi\left(\Sigma\left(Q_m\left(u,\hbar u\right)\right)\right)<\Sigma\left(Q_m\left(u,\hbar u\right)\right).$$

It follows that $\Sigma\left(Q_m\left(u,\hbar u\right)\right)<\Sigma\left(Q_m\left(u,\hbar u\right)\right)$, a contradiction. Therefore, we get $Q_m\left(u,\hbar u\right)=0$, that is $u=\hbar u$ since Q is regular.

Next, we show that u is a unique fixed point of $\hbar$. On the contrary, we suppose that u^* is another fixed point of $\hbar$, i.e., $u^*=\hbar u^*$ and $u\neq u^*$. Then, there exist $\sigma\in S_Q$ such that $(u,\sigma)\in B(H)$ and $(\sigma,u^*)\in B(H)$. Using (iii), we get that $(u,u^*)\in B(\tilde{H})$. Furthermore,

$$0=\frac{1}{2}Q_m\left(u,\hbar u\right)<Q_m\left(u,u^*\right). \quad (32)$$

From the generalized Suzuki-Simulation-$\tilde{H}$-type contractive mapping we have:
$$C_G\leq\zeta\left(\Sigma\left(Q_m\left(\hbar u,\hbar u^*\right)\right),\varphi\left(\Sigma\left(P\left(u,u^*\right)\right)\right)\right)$$
$$\leq\zeta\left(\Sigma\left(Q_m\left(u,u^*\right)\right),\varphi\left(\Sigma\left(P\left(u,u^*\right)\right)\right)\right) \quad (33)$$
$$\leq G\left(\varphi\left(\Sigma\left(P\left(u,u^*\right)\right)\right),\Sigma\left(Q_m\left(\hbar u,\hbar u^*\right)\right)\right).$$

Using Definition 5, we get:
$$\Sigma\left(Q_m\left(u,u^*\right)\right)<\varphi\left(\Sigma\left(P\left(u,u^*\right)\right)\right) \quad (34)$$

where:
$$P\left(u,u^*\right)=\max\left\{Q_m\left(u,u^*\right),Q_m\left(u,\hbar u\right),Q_m\left(u^*,\hbar u^*\right)\right\}$$
$$=\max\left\{Q_m\left(u,u^*\right),0\right\}=Q_m\left(u,u^*\right). \quad (35)$$

From (33)–(35), it follows that:
$$\Sigma\left(Q_m\left(u,u^*\right)\right)<\varphi\left(\Sigma\left(Q_m\left(u,u^*\right)\right)\right)<\Sigma\left(Q_m\left(u,u^*\right)\right).$$

This is an incorrect statement. Hence, $u=u^*$. □

5. Conclusions

First, motivated by [4,10,15], we established a new contractive mapping, which is called the generalized Suzuki-simulation-type contractive mapping. Second, in [19], we constituted a new quasi metric space, which is named the non-Archimedean quasi modular metric space, and so using this, we attained fixed point theorems via generalized Suzuki-simulation-type contractive mapping. Finally, we acquired graphical fixed point results in non-Archimedean quasi modular metric spaces.

Author Contributions: The authours contributed equally in writing this article. Authours read and approved the manuscript.

Funding: This research received no external funding.

Acknowledgments: The authors are grateful to the editor and reviewers for their careful reviews and useful comments.

Conflicts of Interest: The authors declare no conflict of interest.

References

1. Khojasteh, F.; Shukla, S.; Radenovic, S. A new approach to the study of fixed point theorems via simulation functions. *Filomat* **2015**, *29*, 1189–1194. [CrossRef]
2. Argoubi, H.; Samet, B.; Vetro, C. Nonlinear contractions involving simulation functions in a metric space with a partial order. *J. Nonlinear Sci. Appl.* **2015**, *8*, 1082–1094. [CrossRef]
3. Abbas, M.; Latif, A.; Suleiman, Y. Fixed points for cyclic R-contractions and solution of nonlinear Volterra integro-differantial equations. *Fixed Point Theory Appl.* **2016**, *2016*, 61. [CrossRef]
4. Radenovic, S.; Chandok, S. Simulation type functions and coincidence point results. *Filomat* **2018**, *32*, 141–147. [CrossRef]
5. Samet, B. Best proximity point results in partially ordered metric spaces via simulation functions. *Fixed Point Theory Appl.* **2015**, *2015*, 232. [CrossRef]
6. Tchier, F.; Vetro, C.; Vetro, F. Best approximation and variational inequality problems involving a simulation function. *Fixed Point Theory Appl.* **2016**, *2016*, 26. [CrossRef]
7. Gholizadeh, L.; Karapinar, E. Best proximity point results in dislocated metric space via R-functions. *Rev. Real Acad. Cienc. Exactas Físicas Nat. Ser. A Mat.* **2018**, *112*, 1391–1407. [CrossRef]
8. Alsamir, H.; Noorani, M.S.; Shatanawi, W.; Aydi, H.; Akhadkulov, H.; Alanazi, K. Fixed point results in metric-like spaces via σ-simulation functions. *Eur. J. Pure Appl. Math.* **2009**, *12*, 88–100. [CrossRef]
9. Ansari, A.H.; Demma, M.; Guran, L. Fixed point reults for C-class functions in modular spaces. *J. Fixed Point Theory Appl.* **2018**, *20*, 13. [CrossRef]
10. Suzuki, T. A new type of fixed point theorem in metric spaces. *Nonlinear Anal.* **2009**, *71*, 5313–5317. [CrossRef]
11. Hima Bindu, V.M.L.; Kishore, G.N.V.; Rao, K.P.R.; Phani, Y. Suzuki type unique common fixed point theorem in partial metric spaces using (C)-condition. *Math Sci.* **2017**, *11*, 39–45. [CrossRef]
12. Jleli, M.; Samet, B. A new generalization of the Banach contraction principle. *J. Inequal. Appl.* **2014**, *2014*, 38. [CrossRef]
13. Jleli, M.; Karapinar, E.; Samet, B. Further generalization of the Banach contraction principle. *J. Inequal. Appl.* **2014**, *2014*, 439. [CrossRef]
14. Hussain, N.; Parvaneh, V.; Samet, B.; Vetro, C. Some fixed point theorems for generalized contractive mappings in complete metric spaces. *Fixed Point Theory Appl.* **2015**, *2015*, 185. [CrossRef]
15. Liu, X.D.; Chang, S.S.; Xiao, Y.; Zhao, L.C. Existence of fixed points for Θ-type contraction and Θ-type Suzuki contraction in complete metric spaces. *Fixed Point Theory Appl.* **2016**, *2016*, 8. [CrossRef]
16. Ahmad, J.; Al-Mazrooei, A.E.; Cho, Y.J.; Yang, Y. Fixed point results for generalized Θ-contractions. *J. Nonlinear Sci. Appl.* **2017**, *10*, 2350–2358. [CrossRef]
17. Onsod, W.; Kumam, P.; Cho, Y.J. Fixed points of α-Θ-Geraghty type and Θ-Geraghty grahic type contractions. *Appl. Gen. Topol.* **2017**, *18*, 153–171. [CrossRef]
18. Zheng, D.W.; Cai, Z.Y.; Wang, P. New fixed point theorems for α-ψ-contraction in complete metric spaces. *J. Nonlinear Sci. Appl.* **2017**, *10*, 2662–2670. [CrossRef]
19. Girgin, E.; Öztürk, M. (α, β)-ψ-type contraction in non-Archimedean quasi modular metric spaces and applications. *J. Math. Anal.* **2019**, *10*, 19–30.
20. Jachymski, J. The contraction principle for mappings on a metric space with a graph. *Proc. Am. Math. Soc.* **2008**, *136*, 1359–1373. [CrossRef]
21. Öztürk, M.; Abbas, M.; Girgin, E. Common fixed point results of a pair generalized (ψ, φ)-contraction mappings in modular spaces. *Fixed Point Theory Appl.* **2016**, *2016*, 19. [CrossRef]
22. Beg, I.; Butt, A.R.; Radenovic, S. The contraction principle for set value mappings on a metric space with graph. *Comput. Math. Appl.* **2010**, *60*, 1214–1219. [CrossRef]
23. Hussain, N.; Arshad, M.; Shoabid, A. Common fixed point results for α-ψ- contractions on a metric space endowed with a graph. *J. Inequal. Appl.* **2014**, *2014*, 136. [CrossRef]
24. Öztürk, M.; Abbas, M.; Girgin, E. Fixed Points of ψ-Contractive Mappings in Modular Spaces. *Filomat* **2016**, *30*, 3817–3827. [CrossRef]

25. Pansuwan, A.; Sintunavarat, W.; Parvaneh, V.; Cho, Y.J. Some fixed point theorems for (α, θ, k)-contractive multi-valued mappings with some applications. *Fixed Point Theory Appl.* **2015**, *2015*, 132. [CrossRef]
26. Rasham, T.; Shoaib, A.; Alamri, B.A.S.; Arshad, M. Fixed Point Results for Multivalued Contractive Mappings Endowed with Graphic Structure. *J. Math.* **2018**, *2018*, 8. [CrossRef]

© 2019 by the authors. Licensee MDPI, Basel, Switzerland. This article is an open access article distributed under the terms and conditions of the Creative Commons Attribution (CC BY) license (http://creativecommons.org/licenses/by/4.0/).

Article

Ample Spectrum Contractions and Related Fixed Point Theorems

Antonio Francisco Roldán López de Hierro [1],* and Naseer Shahzad [2]

[1] Department of Statistics and Operations Research, University of Granada, 18010 Granada, Spain
[2] Department of Mathematics, Faculty of Science, King Abdulaziz University, P.O.B. 80203, Jeddah 21589, Saudi Arabia; nshahzad@kau.edu.sa
* Correspondence: aroldan@ugr.es

Received: 28 September 2019; Accepted: 28 October 2019; Published: 2 November 2019

Abstract: Simulation functions were introduced by Khojasteh et al. as a method to extend several classes of fixed point theorems by a simple condition. After that, many researchers have amplified the knowledge of such kind of contractions in several ways. R-functions, $(R, \mathcal{S})$-contractions and $(\mathcal{A}, \mathcal{S})$-contractions can be considered as approaches in this direction. A common characteristic of the previous kind of contractive maps is the fact that they are defined by a strict inequality. In this manuscript, we show the advantages of replacing such inequality with a weaker one, involving a family of more general auxiliary functions. As a consequence of our study, we show that not only the above-commented contractions are particular cases, but also another classes of contractive maps correspond to this new point of view.

Keywords: R-function; simulation function; manageable function; fixed point; contractivity condition; binary relation

1. Introduction

Fixed point theory is a branch of mathematics that has multiple applications in almost all scientific fields of study. Mainly, it is used to prove the existence (and, in many cases, also uniqueness) of solutions of great variety of equations arising in theoretical and practical disciplines: matrix equations, differential equations, integral equations, etc. One of its best advantage is the fact that it permits us to deal with linear and nonlinear problems, which makes this discipline into an essential part of nonlinear analysis.

Although it was not the first result in this line of research, Banach contractive mapping principle is widely considered the pioneering statement. Any new result in this area must generalize such principle. There are many directions in which it has been extended and improved: by using weaker contractivity conditions, more general families of auxiliary functions, by involving a partial order, by considering abstract metric spaces, etc.

In recent times, Khojasteh et al. [1] introduced a new class of auxiliary functions, called *simulation functions*, that let us consider a family of contractivity conditions that only involve two arguments: the distance between two points $(d(x,y))$ and the distance between their corresponding images $(d(Tx, Ty))$ under the considered operator. This work quickly attracted the attention of several researchers because of its potential applications (see, for instance, the work of Roldán López de Hierro et al. [2], who slightly modified the original definition, and those of Roldán López de Hierro and Shahzad [3,4], who presented R-functions as extensions of simulation functions).

The above-mentioned classes of contractions have been included in a new family of contractive mappings, called $(\mathcal{A}, \mathcal{S})$-contractions, that extend and unify several results in fixed point theory (see [5]). Theoretical notions introduced in such manuscript were later developed by other researchers (see [6]) even with applications to fuzzy partial differential equations (see [7]) and optimal solutions and applications to nonlinear matrix and integral equations (see [8]). However, in the original definition of $(\mathcal{A}, \mathcal{S})$-contractions, inspired by the previous contributions, the authors established a strict inequality that must be verified for some pairs of points related under a binary relation. In this manuscript, we improve such results in several ways: (1) the given family of auxiliary functions is more general; (2) coherently, the presented contractivity condition is weaker; and (3) the set of points that have to satisfy the contractivity condition is smaller. These improvements let us show that not only the above-commented contractions are particular cases of our study, but also new families of contractive maps correspond to this new approach (see [9–11]). The presented contractions are called *ample spectrum contractions* because they are an attempt to generalize all known contractions that are defined by contractivity conditions that involve only the terms $d(x, y)$ and $d(Tx, Ty)$.

2. Preliminaries

Basic notions and notations for a good understanding of this manuscript are given in [5]. Nevertheless, we recall here the essential facts. Throughout this manuscript, X always stands for a nonempty set. A *binary relation on* X is a nonempty subset $\mathcal{S}$ of the product space $X \times X$. If $(x, y) \in \mathcal{S}$, we denote it by $x\mathcal{S}y$. We write $x\mathcal{S}^*y$ when $x\mathcal{S}y$ and $x \neq y$. Notice that $\mathcal{S}^*$, if it is nonempty, is another binary relation on X. Two points x and y are $\mathcal{S}$-*comparable* if $x\mathcal{S}y$ or $y\mathcal{S}x$. A binary relation $\mathcal{S}$ is:

- *transitive*: If from $x\mathcal{S}y$ and $y\mathcal{S}z$ it follows $x\mathcal{S}z$,
- *reflexive*: If $x\mathcal{S}x$ for each $x \in \mathbb{R}$,
- *antisymmetric*: If from $x\mathcal{S}y$ and $y\mathcal{S}x$ it follows $x = y$.

Reflexive and transitive binary relations are called *preorders* (or *quasiorders*), and, if they are also antisymmetric, then they are *partial orders*. The trivial partial order $\mathcal{S}_X$ is defined by $x\mathcal{S}_X y$ for each $x, y \in X$.

From now on, $\mathbb{N} = \{0, 1, 2, 3, \ldots\}$ stands for the set of all nonnegative integers and $\mathbb{N}^* = \mathbb{N} \setminus \{0\}$. Henceforth, let $T : X \to X$ be a map from X into itself, let (X, d) be a metric space and let $A \subseteq \mathbb{R}$ be a nonempty subset of the set of all real numbers. The range (or image) of d is $\operatorname{ran}(d) = \{d(x, y) : x, y \in X\} \subseteq [0, \infty)$.

If $Tx = x$, then x is a *fixed point of* T. The maps $\{T^n : X \to X\}_{n \in \mathbb{N}}$ defined by $T^0 =$identity, $T^1 = T$ and $T^{n+1} = T \circ T^n$ for all $n \geq 2$ are known as the *iterates* of T. The *Picard sequence of* T based on $x_0 \in X$ is the sequence $\{x_n\}_{n \in \mathbb{N}}$ given by $x_{n+1} = Tx_n$ for all $n \in \mathbb{N}$ (hence, $x_n = T^n x_0$ for each $n \in \mathbb{N}$). When any Picard sequence of T converges to a fixed point of T, we say that T is a *weakly Picard operator*, and if it has a unique fixed point, then T is known as *Picard operator*.

In [5], the authors used the following terminology. Let $\mathcal{S}$ be a binary relation on a metric space (X, d), let $Y \subseteq X$ be a nonempty subset, let $\{x_n\}$ be a sequence in X and let $T : X \to X$ be a self-mapping. We say that:

- A sequence $\{x_n\} \subseteq X$ is *asymptotically regular on* (X, d) if $\{d(x_n, x_{n+1})\} \to 0$.
- T is $\mathcal{S}$-*nondecreasing* if $Tx\mathcal{S}Ty$ for all $x, y \in X$ such that $x\mathcal{S}y$.
- $\{x_n\}$ is $\mathcal{S}$-*nondecreasing* if $x_n \mathcal{S} x_m$ for all $n, m \in \mathbb{N}$ such that $n < m$.
- $\{x_n\}$ is $\mathcal{S}$-*strictly-increasing* if $x_n \mathcal{S}^* x_m$ for all $n, m \in \mathbb{N}$ such that $n < m$.
- T is $\mathcal{S}$-*nondecreasing-continuous* if $\{Tx_n\} \to Tz$ for all $\mathcal{S}$-nondecreasing sequence $\{x_n\} \subseteq X$ such that $\{x_n\} \to z \in X$.
- T is $\mathcal{S}$-*strictly-increasing-continuous* if $\{Tx_n\} \to Tz$ for all $\mathcal{S}$-strictly-increasing sequence $\{x_n\} \subseteq X$ such that $\{x_n\} \to z \in X$.
- Y is $(\mathcal{S}, d)$-*strictly-increasing-complete* if every $\mathcal{S}$-strictly-increasing and d-Cauchy sequence $\{y_n\} \subseteq Y$ is d-convergent to a point of Y.
- Y is $(\mathcal{S}, d)$-*strictly-increasing-precomplete* if there exists a set Z such that $Y \subseteq Z \subseteq X$ and Z is $(\mathcal{S}, d)$-strictly-increasing-complete;
- (X, d) is $\mathcal{S}$-*strictly-increasing-regular* if, for all $\mathcal{S}$-strictly-increasing sequence $\{x_n\} \subseteq X$ such that $\{x_n\} \to z \in X$, it follows that $x_n \mathcal{S} z$ for all $n \in \mathbb{N}$.

We follow the notation given in [12,13]. Next, we list a collection of properties that can be satisfied by a function $\phi : [0, \infty) \to [0, \infty)$.

$(\mathcal{P}_1)$ ϕ is non-decreasing, that is, if $0 \leq t \leq s$, then $\phi(t) \leq \phi(s)$.
$(\mathcal{P}_{10})$ The series $\sum_{n \geq 1} \phi^n(t)$ converges for all $t > 0$.
$(\mathcal{P}_{11})$ $\lim_{n \to \infty} \phi^n(t) = 0$ for all $t > 0$.
$(\mathcal{P}_{12})$ $\phi(t) < t$ for all $t > 0$.
$(\mathcal{P}_{13})$ $\lim_{t \to 0^+} \phi(t) = 0$.
$(\mathcal{P}')$ $\phi(0) = 0$.

It is clear that $(\mathcal{P}_{10}) \Rightarrow (\mathcal{P}_{11})$ and, on the other hand, $(\mathcal{P}_{12}) \Rightarrow (\mathcal{P}_{13})$.

Proposition 1 ([12,13]). *If $(\mathcal{P}_1)$ holds, then $(\mathcal{P}_{10}) \Rightarrow (\mathcal{P}_{11}) \Rightarrow (\mathcal{P}_{12}) \Rightarrow (\mathcal{P}_{13}) \Rightarrow (\mathcal{P}')$.*

Given a function $\alpha : X \times X \to [0, \infty)$, it is possible to redefine the previous notions in terms of α (transitivity, α-admissibility, α-nondecreasing character, α-nondecreasing-continuity, α-strictly-increasing-regularity, (α, d)-strictly-increasing-completeness, (α, d)-strictly-increasing-precompleteness, etc.). For details, see [5]. Such properties can be translated to the previous setting by using the binary relation $\mathcal{S}_\alpha$ on X given, for $x, y \in X$, by

$$x \mathcal{S}_\alpha y \quad \text{if} \quad \alpha(x, y) \geq 1. \tag{1}$$

Lemma 1. *Let (X, d) be a metric space, let $T : X \to X$ be a self-mapping and let $\alpha : X \times X \to [0, \infty)$ be a function. Then, the following properties hold.*

1. *The binary relation $\mathcal{S}_\alpha$ is transitive if, and only if, α is transitive.*
2. *T is α-admissible if, and only if, T is $\mathcal{S}_\alpha$-nondecreasing.*
3. *Given $z_0 \in X$, the mapping T is $(d, \mathcal{S}_\alpha)$-nonincreasing-continuous at z_0 if, and only if, it is (d, α)-right-continuous at z_0.*
4. *T is $(d, \mathcal{S}_\alpha)$-nonincreasing-continuous if, and only if, T is (d, α)-right-continuous.*

In [5], Shahzad et al. introduced the following notions.

Definition 1. *Let $\{a_n\}$ and $\{b_n\}$ be two sequences of real numbers. We say that $\{(a_n, b_n)\}$ is a $(T, \mathcal{S})$-sequence if there exist two sequences $\{x_n\}, \{y_n\} \subseteq X$ such that*

$$x_n \mathcal{S} y_n, \quad a_n = d(T x_n, T y_n) > 0 \quad \text{and} \quad b_n = d(x_n, y_n) > 0 \quad \text{for all } n \in \mathbb{N}.$$

If $\mathcal{S}$ is the trivial binary relation $\mathcal{S}_X$, then $\{(a_n, b_n)\}$ is called a T-sequence.

Remark 1. *Notice that $\{(a_n = d(T x_n, T y_n), b_n = d(x_n, y_n))\}$ is a $(T, \mathcal{S})$-sequence if, and only if,*

$$x_n \mathcal{S}^* y_n \quad \text{and} \quad a_n > 0 \quad \text{for all } n \in \mathbb{N}.$$

Definition 2. *We say that $T : X \to X$ is an $(\mathcal{A}, \mathcal{S})$-contraction if there exists a function $\varrho : A \times A \to \mathbb{R}$ such that T and ϱ satisfy the following four conditions:*

$(\mathcal{A}_1)$ $\operatorname{ran}(d) \subseteq A$.
$(\mathcal{A}_2)$ *If $\{x_n\} \subseteq X$ is a Picard $\mathcal{S}$-nondecreasing sequence of T such that*

$$x_n \neq x_{n+1} \quad \text{and} \quad \varrho(d(x_{n+1}, x_{n+2}), d(x_n, x_{n+1})) > 0 \quad \text{for all } n \in \mathbb{N},$$

then $\{x_n\}$ is asymptotically regular on (X, d) (that is, $\{d(x_n, x_{n+1})\} \to 0$).

$(\mathcal{A}_3)$ If $\{(a_n, b_n)\} \subseteq A \times A$ is a $(T, \mathcal{S})$-sequence such that $\{a_n\}$ and $\{b_n\}$ converge to the same limit $L \geq 0$ and verifying that $L < a_n$ and $\varrho(a_n, b_n) > 0$ for all $n \in \mathbb{N}$, then $L = 0$.
$(\mathcal{A}_4)$ $\varrho(d(Tx, Ty), d(x, y)) > 0$ for all $x, y \in X$ such that $x\mathcal{S}^*y$ and $Tx\mathcal{S}^*Ty$.

In such a case, we say that T is an $(\mathcal{A}, \mathcal{S})$-contraction with respect to ϱ. We denote the family of all $(\mathcal{A}, \mathcal{S})$-contractions from (X, d) into itself with respect to ϱ by $\mathcal{A}_{X,d,\mathcal{S},\varrho,A}$ or, for simplicity, by $\mathcal{A}_\varrho$ when no confusion is possible.

If $\mathcal{S}$ is the trivial binary relation $\mathcal{S}_X$, then T is called an $\mathcal{A}$-contraction (with respect to ϱ).

Condition $(\mathcal{A}_1)$ implies that A is a nonempty set. In some cases, we also consider the following properties.

$(\mathcal{A}'_2)$ If $x_1, x_2 \in X$ are two points such that

$$T^n x_1 \mathcal{S}^* T^n x_2 \quad \text{and} \quad \varrho(d\left(T^{n+1}x_1, T^{n+1}x_2\right), d\left(T^n x_1, T^n x_2\right)) > 0 \quad \text{for all } n \in \mathbb{N},$$

then $\{d(T^n x_1, T^n x_2)\} \to 0$.
$(\mathcal{A}_5)$ If $\{(a_n, b_n)\}$ is a $(T, \mathcal{S})$-sequence such that $\{b_n\} \to 0$ and $\varrho(a_n, b_n) > 0$ for all $n \in \mathbb{N}$, then $\{a_n\} \to 0$.

3. Ample Spectrum Contractions

In this section, we slightly modify the axioms given in [5] in a subtle way in order to consider a wider class of contractions. In what follows, let (X, d) be a metric space, let $\mathcal{S}$ be a binary relation on X and let $T : X \to X$ be a self-mapping.

Definition 3. *Let $\{a_n\}$ and $\{b_n\}$ be two sequences of real numbers. We say that $\{(a_n, b_n)\}$ is a $(T, \mathcal{S}^*)$-sequence if there exist two sequences $\{x_n\}, \{y_n\} \subseteq X$ such that*

$$x_n \mathcal{S}^* y_n, \quad Tx_n \mathcal{S}^* Ty_n, \quad a_n = d(Tx_n, Ty_n) > 0 \quad \text{and} \quad b_n = d(x_n, y_n) > 0 \quad \text{for all } n \in \mathbb{N}.$$

Proposition 2. *Every $(T, \mathcal{S}^*)$-sequence is a $(T, \mathcal{S})$-sequence.*

Definition 4. *We say that $T : X \to X$ is a ample spectrum contraction if there exists a function $\varrho : A \times A \to \mathbb{R}$ such that T and ϱ satisfy the following four conditions:*

$(\mathcal{B}_1)$ A is nonempty and $\{d(x, y) \in [0, \infty) : x, y \in X, x\mathcal{S}^*y\} \subseteq A$.
$(\mathcal{B}_2)$ If $\{x_n\} \subseteq X$ is a Picard $\mathcal{S}$-nondecreasing sequence of T such that

$$x_n \neq x_{n+1} \quad \text{and} \quad \varrho\left(d\left(x_{n+1}, x_{n+2}\right), d\left(x_n, x_{n+1}\right)\right) \geq 0 \quad \text{for all } n \in \mathbb{N},$$

then $\{d(x_n, x_{n+1})\} \to 0$.
$(\mathcal{B}_3)$ If $\{(a_n, b_n)\} \subseteq A \times A$ is a $(T, \mathcal{S}^*)$-sequence such that $\{a_n\}$ and $\{b_n\}$ converge to the same limit $L \geq 0$ and verifying that $L < a_n$ and $\varrho(a_n, b_n) \geq 0$ for all $n \in \mathbb{N}$, then $L = 0$.
$(\mathcal{B}_4)$ $\varrho(d(Tx, Ty), d(x, y)) \geq 0$ for all $x, y \in X$ such that $x\mathcal{S}^*y$ and $Tx\mathcal{S}^*Ty$.

In such a case, we say that T is a ample spectrum contraction with respect to $\mathcal{S}$ and ϱ. We denote the family of all ample spectrum contractions from (X, d) into itself with respect to $\mathcal{S}$ and ϱ by $\mathcal{B}_{X,d,\mathcal{S},\varrho,A}$.

In some cases, we also consider the following properties:

$(\mathcal{B}'_2)$ If $x_1, x_2 \in X$ are two points such that

$$T^n x_1 \mathcal{S}^* T^n x_2 \quad \text{and} \quad \varrho(d\left(T^{n+1}x_1, T^{n+1}x_2\right), d\left(T^n x_1, T^n x_2\right)) \geq 0 \quad \text{for all } n \in \mathbb{N},$$

then $\{d(T^n x_1, T^n x_2)\} \to 0$.

($\mathcal{B}_5$) If $\{(a_n, b_n)\}$ is a $(T, \mathcal{S}^*)$-sequence such that $\{b_n\} \to 0$ and $\varrho(a_n, b_n) \geq 0$ for all $n \in \mathbb{N}$, then $\{a_n\} \to 0$.

Remark 2. *The reader can observe the following facts about the previous assumptions:*

1. *Notice that conditions ($\mathcal{B}_2$), ($\mathcal{B}_3$), ($\mathcal{B}_2'$) and ($\mathcal{B}_5$) establish that, if there exists a sequence (or one point, or two points) verifying some assumptions, then a thesis must hold. However, we point out that, if such kind of sequences (or points) does not exist, then conditions ($\mathcal{B}_2$), ($\mathcal{B}_3$), ($\mathcal{B}_2'$) and ($\mathcal{B}_5$) hold.*
2. *Condition ($\mathcal{B}_2$) follows from ($\mathcal{B}_2'$) using $x_2 = Tx_1$.*
3. *None of the previous conditions establishes a constraint about the values $\{\varrho(0,s) : s \in A\}$ because the first argument is always positive. In fact, it is possible that $0 \notin A$.*
4. *If $x\mathcal{S}^* y$, then $d(x, y) > 0$. Hence, $0 \notin \{d(x,y) \in [0, \infty) : x, y \in X, x\mathcal{S}^* y\}$. Nevertheless, 0 may belong to A.*
5. *If $\mathcal{S}$ is the binary relation such that $x\mathcal{S}y$ if, and only if, $x = y$, then $\{d(x,y) \in [0, \infty) : x, y \in X, x\mathcal{S}^* y\}$ is empty. This is the reason we must impose that A is nonempty.*
6. *Condition ($\mathcal{B}_1$) guarantees that the function ϱ can be applied in the other assumptions. For instance, in ($\mathcal{B}_2$), it is clear that $x_n \mathcal{S}^* x_{n+1}$ and $x_{n+1} \mathcal{S}^* x_{n+2}$ because $\{x_n\}$ is $\mathcal{S}$-nondecreasing and $x_n \neq x_{n+1}$ for all $n \in \mathbb{N}$.*
7. *As the reader can easily check in the proofs of the following results, we could also have supposed in Condition ($\mathcal{B}_3$) that $\{x_n\}$ and $\{y_n\}$ are appropriate subsequences of the same Picard sequence $\{z_n = T^n z_0\} \subseteq X$ (in the sense that $x_n = z_{p(n)}$ and $y_n = z_{q(n)}$ being $n \leq p(n) < q(n)$ for all $n \in \mathbb{N}$). In order not to complicate the proofs, we do not include such assumption.*

Proposition 3. *If $\varrho(t, s) \leq s - t$ for all $t, s \in A \cap (0, \infty)$, then ($\mathcal{B}_5$) holds.*

Proof. Assume that $\{a_n\}, \{b_n\} \subset (0, \infty) \cap A$ are two sequences such that $\{b_n\} \to 0$ and $\varrho(a_n, b_n) \geq 0$ for all $n \in \mathbb{N}$. Since $a_n, b_n \in (0, \infty) \cap A$, then $0 < \varrho(a_n, b_n) \leq b_n - a_n$ for all $n \in \mathbb{N}$. As a consequence, $0 < a_n \leq b_n$ for all $n \in \mathbb{N}$, which means that $\{a_n\} \to 0$. □

The previous definition generalizes the notion of $(\mathcal{A}, \mathcal{S})$-contraction, as we prove in the following result:

Theorem 1. *Every $(\mathcal{A}, \mathcal{S})$-contraction is an ample spectrum contraction (with respect to the same function ϱ). Furthermore, if it satisfies ($\mathcal{A}_2'$) (respectively, ($\mathcal{A}_5$)), then it also verifies ($\mathcal{B}_2'$) (respectively, ($\mathcal{B}_5$)).*

In particular, we prove the following implications:

$$(\mathcal{A}_1) \Rightarrow (\mathcal{B}_1),$$
$$(\mathcal{A}_4) \Rightarrow (\mathcal{B}_4),$$
$$(\mathcal{A}_2) + (\mathcal{A}_4) \Rightarrow (\mathcal{B}_2),$$
$$(\mathcal{A}_3) + (\mathcal{A}_4) \Rightarrow (\mathcal{B}_3),$$
$$(\mathcal{A}_4) + (\mathcal{A}_5) \Rightarrow (\mathcal{B}_5),$$
$$(\mathcal{A}_2') + (\mathcal{A}_4) \Rightarrow (\mathcal{B}_2').$$

Proof. Let (X, d) be a metric space, let $T : X \to X$ be a mapping and let $\varrho : A \times A \to \mathbb{R}$ be a function. Clearly, $(\mathcal{A}_1) \Rightarrow (\mathcal{B}_1)$ and $(\mathcal{A}_4) \Rightarrow (\mathcal{B}_4)$. Next, we prove the rest of conditions.

$[(\mathcal{A}_2') + (\mathcal{A}_4) \Rightarrow (\mathcal{B}_2')]$ Let $x_1, x_2 \in X$ be two points such that

$$T^n x_1 \mathcal{S}^* T^n x_2 \quad \text{and} \quad \varrho(d\left(T^{n+1} x_1, T^{n+1} x_2\right), d\left(T^n x_1, T^n x_2\right)) \geq 0 \quad \text{for all } n \in \mathbb{N}.$$

Let us denote

$$x_n^1 = T^n x_1 \quad \text{and} \quad x_n^2 = T^n x_2 \quad \text{for all } n \in \mathbb{N}.$$

Hence, by hypothesis, $x_n^1 = T^n x_1 \mathcal{S}^* T^n x_2 = x_n^2$ and $Tx_n^1 = T^{n+1} x_1 \mathcal{S}^* T^{n+1} x_2 = Tx_n^2$. Applying Condition $(\mathcal{A}_4)$, for all $n \in \mathbb{N}$,

$$\varrho(d\left(T^{n+1}x_1, T^{n+1}x_2\right), d\left(T^n x_1, T^n x_2\right)) = \varrho(d\left(Tx_n^1, Tx_n^2\right), d\left(x_n^1, x_n^2\right)) > 0.$$

Therefore, Condition $(\mathcal{A}_2')$ implies that $\{d(T^n x_1, T^n x_2)\} \to 0$.

$[(\mathcal{A}_2) + (\mathcal{A}_4) \Rightarrow (\mathcal{B}_2)]$ It follows as in the previous implication by using $x_1 = x_0$ and $x_2 = Tx_0$.

$[(\mathcal{A}_3) + (\mathcal{A}_4) \Rightarrow (\mathcal{B}_3)]$ Let $\{(a_n, b_n)\} \subseteq A \times A$ be a $(T, \mathcal{S}^*)$-sequence such that $\{a_n\}$ and $\{b_n\}$ converge to the same limit $L \geq 0$ and verifying that $L < a_n$ and $\varrho(a_n, b_n) \geq 0$ for all $n \in \mathbb{N}$. By definition, there are two sequences $\{x_n\}, \{y_n\} \subseteq X$ such that

$$x_n \mathcal{S}^* y_n, \quad Tx_n \mathcal{S}^* Ty_n, \quad a_n = d(Tx_n, Ty_n) > 0 \quad \text{and} \quad b_n = d(x_n, y_n) > 0 \quad \text{for all } n \in \mathbb{N}.$$

As $x_n \mathcal{S}^* y_n$ and $Tx_n \mathcal{S}^* Ty_n$, then it follows from $(\mathcal{A}_4)$ that

$$\varrho(a_n, b_n) = \varrho(d(Tx_n, Ty_n), d(x_n, y_n)) > 0 \quad \text{for all } n \in \mathbb{N}.$$

Therefore, applying $(\mathcal{A}_3)$, we conclude that $L = 0$.

$[(\mathcal{A}_4) + (\mathcal{A}_5) \Rightarrow (\mathcal{B}_5)]$ Let $\{(a_n, b_n)\}$ be a $(T, \mathcal{S}^*)$-sequence such that $\{b_n\} \to 0$ and $\varrho(a_n, b_n) \geq 0$ for all $n \in \mathbb{N}$. By definition, there exist two sequences $\{x_n\}, \{y_n\} \subseteq X$ such that

$$x_n \mathcal{S}^* y_n, \quad Tx_n \mathcal{S}^* Ty_n, \quad a_n = d(Tx_n, Ty_n) > 0 \quad \text{and} \quad b_n = d(x_n, y_n) > 0 \quad \text{for all } n \in \mathbb{N}.$$

As $x_n \mathcal{S}^* y_n$ and $Tx_n \mathcal{S}^* Ty_n$, then it follows from $(\mathcal{A}_4)$ that

$$\varrho(a_n, b_n) = \varrho(d(Tx_n, Ty_n), d(x_n, y_n)) > 0 \quad \text{for all } n \in \mathbb{N}.$$

Therefore, applying $(\mathcal{A}_5)$, we conclude that $\{a_n\} \to 0$. □

The previous theorem provides us a large list of ample spectrum contractions because every $(\mathcal{A}, \mathcal{S})$-contraction is an ample spectrum contraction. In particular, as the authors proved in [3,5], the following ones are examples of ample spectrum contractions:

- Banach contractions;
- Meir–Keeler contractions (see [14,15]);
- $\mathcal{Z}$-contractions involving *simulation functions* (see [1,2]);
- manageable contractions (see [16]);
- Geraghty contractions (see [17]); and
- R-contractions (see [3,5]).

The converse of Theorem 1 is false, as we show in the following example:

Example 1. *Let $X = \{0, 1, 3\}$ be endowed with the Euclidean metric $d_E(x, y) = |x - y|$ and the usual order $\leq$. Hence, (X, d_E) is a complete metric space. Let $A = \operatorname{ran}(d_E) = \{0, 1, 2, 3\}$ and let $T : X \to X$ and $\varrho : A \times A \to \mathbb{R}$ be defined by*

$$Tx = \begin{cases} 0, & \text{if } x \in \{0, 1\}, \\ 1, & \text{if } x = 3; \end{cases} \qquad \varrho(t, s) = 0 \text{ for all } t, s \in A.$$

Then, T is not an $(\mathcal{A}, \leq)$-contraction with respect to ϱ because, if $x = 1$ and $y = 3$, then $x < y$ and $Tx < Ty$, but $\varrho(d(Tx, Ty), d(x, y)) = 0$. Let us show that T is an ample spectrum contraction with respect to ϱ and $\leq$. Condition $(\mathcal{B}_4)$ is obvious. Properties $(\mathcal{B}_2)$ and $(\mathcal{B}_2')$ follows from the fact that any Picard sequence $\{x_n\}$ of T must verify $x_n = 0$ for all $n \geq 3$. Taking into account that any convergent sequence on A is almost constant (because it is discrete), Axioms $(\mathcal{B}_3)$ and $(\mathcal{B}_5)$ are satisfied because such kind of sequences do not exist. Hence, T is an ample spectrum contraction with respect to ϱ and $\leq$.

The notion of $(T, \mathcal{S}^*)$-sequence plays a key role in the definition of ample spectrum contraction. In fact, if we had not changed the notion of $(T, \mathcal{S})$-sequence by the concept of $(T, \mathcal{S}^*)$-sequence in Definition 4, then there would have not been any relationship between $(\mathcal{A}, \mathcal{S})$-contractions and ample spectrum contractions. We illustrate this affirmation with the following example.

Example 2. Let $X = [0,2] \cup C \cup D$, where $C = \{ 10m \in \mathbb{N} : m \in \mathbb{N}^* \}$ and $D = \{ 10m + 4 \in \mathbb{N} : m \in \mathbb{N}^* \}$. Assume that X is endowed with the Euclidean metric $d_E(x,y) = |x - y|$ and the usual order $\leq$. Hence, (X, d_E) is a complete metric space. The range of d_E can be expressed as

$$\mathrm{ran}(d_E) = [0,2] \cup \{4\} \cup B \quad \text{where } B \subset [6, \infty).$$

Let $A = \mathrm{ran}(d_E)$ and let $T : X \to X$ and $\varrho : A \times A \to \mathbb{R}$ be defined by

$$Tx = \begin{cases} \dfrac{x}{4}, & \text{if } x \in [0,2], \\ 1 + \dfrac{1}{m}, & \text{if } x = 10m \in C \quad (\text{for some } m \in \mathbb{N}^*), \\ 0, & \text{if } x = 10m + 4 \in D \quad (\text{for some } m \in \mathbb{N}^*); \end{cases}$$

$$\varrho(t,s) = \begin{cases} 0, & \text{if } t > 1 \text{ and } s \geq 1, \\ \dfrac{s}{2} - t, & \text{otherwise.} \end{cases}$$

Notice that T satisfies the following properties.

(p_1) $T(X) \subset [0,2]$. In particular, $|Tx - Ty| \leq 2$ for all $x, y \in X$.
(p_2) If $x, y \in X$ are two different points such that $x \in C \cup D$ or $y \in C \cup D$, then $|x - y| \geq 4$. In particular, if $|x - y| < 4$, then $x, y \in [0,2]$.
(p_3) For all $x_0 \in X$, the Picard sequence of T based on x_0 verifies $x_{n+1} = \dfrac{Tx_0}{4^n}$ for all $n \in \mathbb{N}$. Thus, every Picard sequence of T converges to zero.

Let us show that T is an ample spectrum contraction with respect to ϱ and $\leq$.
($\mathcal{B}_2$) Let $\{x_n\} \subseteq X$ be a Picard $\mathcal{S}$-nondecreasing sequence of T such that

$$x_n \neq x_{n+1} \quad \text{and} \quad \varrho\left(d_E(x_{n+1}, x_{n+2}), d_E(x_n, x_{n+1})\right) \geq 0 \quad \text{for all } n \in \mathbb{N}.$$

Since $\{x_n\} \to 0$, $\{d_E(x_n, x_{n+1})\} \to 0$.
($\mathcal{B}_3$) Let $\{(a_n, b_n)\} \subseteq A \times A$ be a $(T, <)$-sequence such that $\{a_n\}$ and $\{b_n\}$ converge to the same limit $L \geq 0$ and verifying that $L < a_n$ and $\varrho(a_n, b_n) \geq 0$ for all $n \in \mathbb{N}$. By definition, there are two sequences $\{x_n\}, \{y_n\} \subseteq X$ such that

$$x_n < y_n, \quad Tx_n < Ty_n, \quad a_n = d_E(Tx_n, Ty_n) > 0 \quad \text{and} \quad b_n = d(x_n, y_n) > 0 \quad \text{for all } n \in \mathbb{N}.$$

As $a_n = d_E(Tx_n, Ty_n) \in [0,2]$, then $L \leq 2$. Since $\{b_n = d_E(x_n, y_n)\} \to L \leq 2$, there exists $n_0 \in \mathbb{N}$ such that $d_E(x_n, y_n) < 4$ for all $n \geq n_0$. By (p_2), we have that $x_n, y_n \in [0,2]$ for all $n \geq n_0$. Therefore, for all $n \geq n_0$,

$$a_n = d_E(Tx_n, Ty_n) = \left| \frac{x_n}{4} - \frac{y_n}{4} \right| = \frac{|x_n - y_n|}{4} = \frac{b_n}{4}.$$

Letting $n \to \infty$, we deduce that $L = L/4$, so $L = 0$.
($\mathcal{B}_4$) Let $x, y \in X$ be two points such that $x < y$ and $Tx < Ty$. To prove that $\varrho\left(d(Tx, Ty), d(x,y)\right) \geq 0$, we observe three cases.

▶ If $\varrho\left(d(Tx, Ty), d(x,y)\right) = 0$, then ($\mathcal{B}_4$) holds. Hence, in what follows, we can assume that

$$\varrho\left(d(Tx, Ty), d(x,y)\right) = \frac{|x - y|}{2} - |Tx - Ty| = \frac{y - x}{2} - (Ty - Tx),$$

which corresponds to the case in which $|Tx - Ty| \leq 1$ or $|x - y| < 1$.

▶ If $|x-y| \geq 4$, then, by (p_1),

$$\varrho\left(d(Tx,Ty),d(x,y)\right) = \frac{|x-y|}{2} - |Tx-Ty| \geq \frac{4}{2} - 2 = 0.$$

▶ On the contrary case, if $|x-y| < 4$, then x or y cannot belong to $C \cup D$. Then, necessarily, $x,y \in [0,2]$, thus

$$\varrho\left(d(Tx,Ty),d(x,y)\right) = \frac{|x-y|}{2} - |Tx-Ty| = \frac{|x-y|}{2} - \left|\frac{x}{4} - \frac{y}{4}\right| = \frac{|x-y|}{4} > 0,$$

which means that $(\mathcal{B}_4)$ holds.

In any case, $(\mathcal{B}_4)$ holds.

The following result is useful in order to study when an ample spectrum contraction can have multiple fixed points.

Proposition 4. Let (X,d) be a metric space endowed with a binary relation $\mathcal{S}$ and let $T : X \to X$ and $\varrho : A \times A \to \mathbb{R}$ be two maps such that $(\mathcal{B}_1)$, $(\mathcal{B}'_2)$ and $(\mathcal{B}_4)$ holds. If $\omega, \omega' \in X$ are two $\mathcal{S}$-comparable fixed points of T, then $\omega = \omega'$.

Proof. Reasoning by contradiction, assume that ω and ω' are two distinct fixed points of T. As ω and ω' are $\mathcal{S}$-comparable, we can suppose, without loss of generality, that $\omega \mathcal{S} \omega'$. Hence, $\omega \mathcal{S}^* \omega'$ and also $T\omega \mathcal{S}^* T\omega'$. Let $a_n = d(\omega, \omega') > 0$ for all $n \in \mathbb{N}$. By using $(\mathcal{B}_4)$, for all $n \in \mathbb{N}$,

$$\varrho(a_{n+1}, a_n) = \varrho\left(d(\omega,\omega'), d(\omega,\omega')\right) = \varrho\left(d(T\omega, T\omega'), d(\omega,\omega')\right) \geq 0.$$

Therefore, it follows from $(\mathcal{B}'_2)$ that $\{a_n = d(\omega,\omega')\} \to 0$, which is a contradiction. Thus, $\omega = \omega'$. □

4. Fixed Point Theorems Involving Ample Spectrum Contractions

Once we have changed the notions of $(T,\mathcal{S})$-sequence and $(\mathcal{A},\mathcal{S})$-contraction by the concepts of $(T,\mathcal{S}^*)$-sequence and ample spectrum contraction, we are ready to introduce the main results of the manuscript, which is the aim of the current section. Concretely, as we show below, the following one is the most general theorem of this manuscript.

Theorem 2. Let (X,d) be a metric space endowed with a transitive binary relation $\mathcal{S}$ and let $T : X \to X$ be an $\mathcal{S}$-nondecreasing ample spectrum contraction with respect to $\varrho : A \times A \to \mathbb{R}$. Suppose that $T(X)$ is $(\mathcal{S},d)$-strictly-increasing-precomplete and there exists a point $x_0 \in X$ such that $x_0 \mathcal{S} T x_0$. Assume that at least one of the following conditions is fulfilled:

(a) T is $\mathcal{S}$-strictly-increasing-continuous.
(b) (X,d) is $\mathcal{S}$-strictly-increasing-regular and Condition $(\mathcal{B}_5)$ holds.
(c) (X,d) is $\mathcal{S}$-strictly-increasing-regular and $\varrho(t,s) \leq s - t$ for all $t,s \in A \cap (0,\infty)$.

Then, the Picard sequence of T based on x_0 converges to a fixed point of T. In particular, T has at least a fixed point.

Notice that the metric space (X,d) needs not to be complete.

Proof. Let $x_0 \in X$ be a point such that $x_0 \mathcal{S} T x_0$ and let $\{x_{n+1} = Tx_n\}_{n \geq 0}$ be the Picard sequence of T based on x_0. If there exists some $n_0 \in \mathbb{N}$ such that $x_{n_0+1} = x_{n_0}$, then x_{n_0} is a fixed point of T, and $\{x_n\}$ converges to such point. On the contrary case, assume that $x_n \neq x_{n+1}$ for all $n \in \mathbb{N}$. As T is $\mathcal{S}$-nondecreasing and $x_0 \mathcal{S} T x_0 = x_1$, then $x_n \mathcal{S} x_{n+1}$ for all $n \in \mathbb{N}$, and, as $\mathcal{S}$ is transitive,

$$x_n \mathcal{S} x_m \quad \text{for all } n, m \in \mathbb{N} \text{ such that } n < m. \tag{2}$$

In fact, as $x_n \neq x_{n+1}$ for all $n \in \mathbb{N}$, then

$$x_n \mathcal{S}^* x_{n+1} \quad \text{and} \quad T x_n \mathcal{S}^* T x_{n+1} \quad \text{for all } n \in \mathbb{N}. \tag{3}$$

Let consider the sequence $\{d(x_n, x_{n+1})\} \subseteq A$. Taking into account Equation (3) and the fact that T is an ample spectrum contraction, Condition ($\mathcal{B}_4$) implies that, for all $n \in \mathbb{N}$,

$$\varrho(d\left(T^{n+1} x_0, T^{n+2} x_0\right), d\left(T^n x_0, T^{n+1} x_0\right)) = \varrho\left(d(T x_n, T x_{n+1}), d(x_n, x_{n+1})\right) \geq 0.$$

Applying ($\mathcal{B}_2$), we deduce that $\{x_n = T^n x_0\}$ is an asymptotically regular sequence on (X, d), that is, $\{d(x_n, x_{n+1})\} \to 0$.

Let us show that $\{x_n\}$ is an $\mathcal{S}$-strictly-increasing sequence. Indeed, in view of Equation (2), assume that there exists $n_0, m_0 \in \mathbb{N}$ such that $n_0 < m_0$ and $x_{n_0} = x_{m_0}$. If $p_0 = m_0 - n_0 \in \mathbb{N} \backslash \{0\}$, then $x_{n_0} = x_{n_0 + k p_0}$ for all $k \in \mathbb{N}$. In particular, the sequence $\{d(x_n, x_{n+1})\}$ contains the constant subsequence

$$\left\{ d(x_{n_0 + k p_0}, x_{n_0 + k p_0 + 1}) = d(x_{n_0}, x_{n_0 + 1}) > 0 \right\}_{k \in \mathbb{N}},$$

which contradicts the fact that $\{d(x_n, x_{n+1})\} \to 0$. This contradiction guarantees that $x_n \neq x_m$ for all $n \neq m$, thus $x_n \mathcal{S}^* x_m$ for all $n, m \in \mathbb{N}$ such that $n < m$, that is, $\{x_n\}$ is an $\mathcal{S}$-strictly-increasing sequence.

Next, we show that $\{x_n\}$ is a Cauchy sequence reasoning by contradiction. If $\{x_n\}$ is not a Cauchy sequence, then there exist $\varepsilon_0 > 0$ and two subsequences $\{x_{n(k)}\}$ and $\{x_{m(k)}\}$ of $\{x_n\}$ such that

$$k \leq n(k) < m(k), \quad d(x_{n(k)}, x_{m(k)-1}) \leq \varepsilon_0 < d(x_{n(k)}, x_{m(k)}) \quad \text{for all } k \in \mathbb{N},$$

$$\lim_{k \to \infty} d(x_{n(k)}, x_{m(k)}) = \lim_{k \to \infty} d(x_{n(k)-1}, x_{m(k)-1}) = \varepsilon_0.$$

Let $L = \varepsilon_0 > 0$, $\{a_k = d(x_{n(k)}, x_{m(k)})\} \to L$ and $\{b_k = d(x_{n(k)-1}, x_{m(k)-1})\} \to L$. As $n(k) < m(k)$ (and $n(k) - 1 < m(k) - 1$), then $x_{n(k)} \mathcal{S}^* x_{m(k)}$ and $x_{n(k)-1} \mathcal{S}^* x_{m(k)-1}$. Thus, $\{(a_k, b_k)\}$ is a $(T, \mathcal{S}^*)$-sequence. Since $L = \varepsilon_0 < d(x_{n(k)}, x_{m(k)}) = a_k$ and

$$\varrho(a_k, b_k) = \varrho\left(d(x_{n(k)}, x_{m(k)}), d(x_{n(k)-1}, x_{m(k)-1})\right)$$
$$= \varrho\left(d(T x_{n(k)-1}, T x_{m(k)-1}), d(x_{n(k)-1}, x_{m(k)-1})\right) \geq 0$$

for all $k \in \mathbb{N}$, Condition ($\mathcal{B}_3$) guarantees that $\varepsilon_0 = L = 0$, which is a contradiction. As a consequence, $\{x_n\}$ is a Cauchy sequence. Since $\{x_n\}_{n \geq 1} \subseteq T(X)$ and $T(X)$ is $(\mathcal{S}, d)$-strictly-increasing-precomplete, there is a subset $Z \subseteq X$ such that $T(X) \subseteq Z \subseteq X$ and Z is $(\mathcal{S}, d)$-strictly-increasing-complete. In particular, as $\{x_n\}$ is an $\mathcal{S}$-strictly-increasing and Cauchy sequence, there exists $z \in Z \subseteq X$ such that $\{x_n\} \to z$. Let us show that z is a fixed point of T considering three cases.

Case 1. Assume that T is $\mathcal{S}$-strictly-increasing-continuous. In this case, $\{x_{n+1} = T x_n\} \to T z$, so $T z = z$.

Case 2. Assume that (X, d) is $\mathcal{S}$-strictly-increasing-regular and condition ($\mathcal{B}_5$) holds. In this case, as $\{x_n\}$ is an $\mathcal{S}$-strictly-increasing sequence such that $\{x_n\} \to z$, it follows that

$$x_n \mathcal{S} z \text{ for all } n \in \mathbb{N}. \tag{4}$$

Since T is $\mathcal{S}$-nondecreasing,
$$Tx_n \mathcal{S} Tz \quad \text{for all } n \in \mathbb{N}. \tag{5}$$

Let $a_n = d(x_{n+1}, Tz) = d(Tx_n, Tz)$ and $b_n = d(x_n, z)$ for all $n \in \mathbb{N}$. Clearly, $\{b_n\} \to 0$. Notice that
$$b_n = 0 \;\Rightarrow\; a_n = 0 \tag{6}$$

because
$$b_n = 0 \;\Leftrightarrow\; x_n = z \;\Rightarrow\; x_{n+1} = Tx_n = Tz \;\Leftrightarrow\; a_n = 0.$$

Let consider the set
$$\Omega = \{n \in \mathbb{N} : a_n = 0\} = \{n \in \mathbb{N} : d(x_{n+1}, Tz) = 0\}.$$

Subcase 2.1. Assume that Ω is finite. In this case, there exists $n_0 \in \mathbb{N}$ such that $d(x_{n+1}, Tz) = a_n > 0$ for all $n \geq n_0$. By (6), $d(x_n, z) = b_n > 0$ for all $n \geq n_0$. In this case, $\{(a_n, b_n)\}_{n \geq n_0}$ is a $(T, \mathcal{S})$-sequence (because $a_n = d(Tx_n, Tz) > 0$ and $b_n = d(x_n, z) > 0$ for all $n \geq n_0$). In particular, $x_n \neq z$ and $Tx_n \neq Tz$ for all $n \geq n_0$. By Equations (4) and (5), we deduce that $x_n \mathcal{S}^* z$ and $Tx_n \mathcal{S}^* Tz$ for all $n \geq n_0$. It follows from $(\mathcal{B}_4)$ that
$$\varrho(a_n, b_n) = \varrho\left(d(Tx_n, Tz), d(x_n, z)\right) \geq 0 \quad \text{for all } n \geq n_0.$$

As a consequence, as $(\mathcal{B}_5)$ holds, we conclude that $\{a_n = d(x_{n+1}, Tz)\} \to 0$, that is, $\{x_{n+1}\} \to Tz$, which guarantees that $Tz = z$.

Subcase 2.2. Assume that Ω is not finite. In this case, there exists a subsequence $\{x_{n(k)}\}$ of $\{x_n\}$ such that
$$d(x_{n(k)+1}, Tz) = 0 \quad \text{for all } k \in \mathbb{N}.$$

Hence, $x_{n(k)+1} = Tz$ for all $k \in \mathbb{N}$. Since $\{x_n\} \to z$ and $\left\{x_{n(k)+1}\right\} \to Tz$, $Tz = z$.

Case 3. Assume that (X, d) is $\mathcal{S}$-strictly-increasing-regular and $\varrho(t, s) \leq s - t$ for all $t, s \in A \cap (0, \infty)$. Proposition 3 guarantees that Item (b) is applicable.

In any case, we conclude that z is a fixed point of T. □

In the following result, we describe sufficient conditions in order to guarantee uniqueness of the fixed point.

Theorem 3. *Under the hypotheses of Theorem 2, assume that the following properties are fulfilled:*

▶ *Condition $(\mathcal{B}'_2)$ holds; and*
▶ *for all $x, y \in \text{Fix}(T)$, there exists $z \in X$ such that z is, at the same time, $\mathcal{S}$-comparable to x and $\mathcal{S}$-comparable to y.*

Then, T has a unique fixed point.

Proof. Let $x, y \in \text{Fix}(T)$ be two fixed points of T. By hypothesis, there exists $z_0 \in X$ such that z_0 is, at the same time, $\mathcal{S}$-comparable to x and $\mathcal{S}$-comparable to y. Let $\{z_n\}$ be the Picard sequence of T based on z_0, that is, $z_{n+1} = Tz_n$ for all $n \in \mathbb{N}$. We prove that $x = y$ by showing that $\{z_n\} \to x$ and $\{z_n\} \to y$. We first use x, but the same argument is valid for y.

Since z_0 is $\mathcal{S}$-comparable to x, assume that $z_0 \mathcal{S} x$ (the case $x \mathcal{S} z_0$ is similar). As T is $\mathcal{S}$-nondecreasing, $z_n \mathcal{S} x$ for all $n \in \mathbb{N}$. If there exists $n_0 \in \mathbb{N}$ such that $z_{n_0} = x$, then $z_n = x$ for all $n \geq n_0$. In particular, $\{z_n\} \to x$ and the proof is finished. On the contrary case, assume that $z_n \neq x$ for all $n \in \mathbb{N}$. Therefore $z_n \mathcal{S}^* x$ and $Tz_n \mathcal{S}^* Tx$ for all $n \in \mathbb{N}$. Using the contractivity Condition $(\mathcal{B}_4)$, for all $n \in \mathbb{N}$,
$$0 \leq \varrho(d(Tz_n, Tx), d(z_n, x)) = \varrho(d(T^{n+1}z_0, T^{n+1}x), d(T^n z_0, T^n x)).$$

It follows from $(\mathcal{B}'_2)$ that $\{d(T^n z_0, T^n x)\} \to 0$, that is, $\{z_n\} \to x$. □

5. Consequences

In this section, we illustrate how many well known theorems in fixed point theory (that involve only $d(x,y)$ and $d(Tx,Ty)$ in their contractivity conditions) can be deduced from our main results.

5.1. Meir–Keeler Contractions

Meir and Keeler generalized the Banach theorem in a way that have attracted much attention in the last 40 years.

Definition 5 (Meir and Keeler [15]). *A Meir–Keeler contraction is a mapping $T : X \to X$ from a metric space (X,d) into itself such that for all $\varepsilon > 0$, there exists $\delta > 0$ verifying that if $x, y \in X$ and $\varepsilon \leq d(x,y) < \varepsilon + \delta$, then $d(Tx, Ty) < \varepsilon$.*

Lim characterized this kind of mappings in terms of a contractivity condition using the following class of auxiliary functions.

Definition 6 (Lim [14]). *A function $\phi : [0, \infty) \to [0, \infty)$ is called an L-function if*

(a) $\phi(0) = 0$;
(b) $\phi(t) > 0$ for all $t > 0$; and
(c) for all $\varepsilon > 0$, there exists $\delta > 0$ such that $phi(t) \leq \varepsilon$ for all $t \in [\varepsilon, \varepsilon + \delta]$.

Each L-function must satisfy:

$$\phi(t) \leq t \quad \text{for all } t \in [0, \infty). \tag{7}$$

Theorem 4 (Lim [14], Theorem 1). *Let (X,d) be a metric space and let $T : X \to X$ be a self-mapping. Then, T is a Meir–Keeler mapping if, and only if, there exists an (non-decreasing, right-continuous) L-map ϕ such that*

$$d(Tx, Ty) < \phi(d(x,y)) \quad \text{for all } x, y \in X \text{ verifying } d(x,y) > 0. \tag{8}$$

Meir and Keeler [15] demonstrated the following fixed point theorem by using a result of Chu and Diaz [18].

Theorem 5 (Meir and Keeler [15]). *Every Meir–Keeler contraction from a complete metric space into itself has a unique fixed point.*

We prove that this result can be immediately deduced from our main statements.

Theorem 6. *Every Meir–Keeler contraction is an ample spectrum contraction that also verifies $(\mathcal{B}'_2)$ and $(\mathcal{B}_5)$.*

Proof. Let (X,d) be a metric space and let $T : X \to X$ be a Meir–Keeler contraction. By Theorem 4, there exists an L-map $\phi : [0, \infty) \to [0, \infty)$ verifying Equation (8). Let $A = \text{ran}(d)$ and let define $\varrho_\phi : A \times A \to \mathbb{R}$ by $\varrho_\phi(t,s) = \phi(s) - t$ for all $t, s \in A$. Let us show that T is an ample spectrum contraction with respect to ϱ_ϕ.

$(\mathcal{B}'_2)$ Let $x_1, x_2 \in X$ be two points such that

$$T^n x_1 \neq T^n x_2 \quad \text{and} \quad \varrho_\phi(d\left(T^{n+1}x_1, T^{n+1}x_2\right), d\left(T^n x_1, T^n x_2\right)) \geq 0 \quad \text{for all } n \in \mathbb{N}.$$

As $d(T^n x_1, T^n x_2) > 0$, it follows from Equations (7) and (8) that, for all $n \in \mathbb{N}$,

$$d\left(T^{n+1}x_1, T^{n+1}x_2\right) = d(TT^n x_1, TT^n x_2) < \phi(d(T^n x_1, T^n x_2)) \leq d(T^n x_1, T^n x_2).$$

As $\{d(T^n x_1, T^n x_2)\}$ is a bounded-below decreasing sequence of real numbers, it is convergent. Let $L \geq 0$ be its limit. To prove that $L = 0$, we reason by contradiction. Assume that $L > 0$. Hence,

$$0 < L \leq d(T^{n+1}x_1, T^{n+1}x_2) < \phi(d(T^n x_1, T^n x_2)) \leq d(T^n x_1, T^n x_2) \quad \text{for all } n \in \mathbb{N}.$$

Letting $\varepsilon = L > 0$ in Condition (c) of Definition 6, there exists $\delta > 0$ such that $\phi(t) \leq \varepsilon = L$ for all $t \in [\varepsilon, \varepsilon + \delta]$. As $\{d(T^n x_1, T^n x_2)\} \searrow L^+$, there exists $n_0 \in \mathbb{N}$ such that $L < d(T^{n_0} x_1, T^{n_0} x_2) < L + \delta$ for all $n \geq n_0$. Therefore,

$$\phi(d(T^{n_0} x_1, T^{n_0} x_2)) \leq \varepsilon = L < \phi(d(T^{n_0} x_1, T^{n_0} x_2)),$$

which is a contradiction. Thus, $L = 0$ and $\{d(T^n x_1, T^n x_2)\} \to 0$.

($\mathcal{B}_2$) It follows from ($\mathcal{B}'_2$).

($\mathcal{B}_3$) Let $\{(a_n, b_n)\} \subseteq A \times A$ be a T-sequence such that $\{a_n\}$ and $\{b_n\}$ converge to the same limit $L \geq 0$ and verifying that $L < a_n$ and $\varrho_\phi(a_n, b_n) \geq 0$ for all $n \in \mathbb{N}$. By definition, there exist two sequences $\{x_n\}, \{y_n\} \subseteq X$ such that

$$a_n = d(Tx_n, Ty_n) > 0 \quad \text{and} \quad b_n = d(x_n, y_n) > 0 \quad \text{for all } n \in \mathbb{N}.$$

Notice that, from Equation (8), for all $n \in \mathbb{N}$,

$$L < a_n = d(Tx_n, Ty_n) < \phi(d(x_n, y_n)) = \phi(b_n) \leq b_n.$$

To prove that $L = 0$, assume that $L > 0$. Letting $\varepsilon = L > 0$ in Condition (c) of Definition 6, there exists $\delta > 0$ such that

$$\phi(t) \leq \varepsilon = L \quad \text{for all } t \in [\varepsilon, \varepsilon + \delta].$$

As $\{d(x_n, y_n)\} \searrow L^+$, there exists $n_0 \in \mathbb{N}$ such that $L < d(x_n, y_n) < L + \delta$ for all $n \geq n_0$. Therefore,

$$\phi(d(x_{n_0}, y_{n_0})) \leq \varepsilon = L < \phi(d(x_{n_0}, y_{n_0})),$$

which is a contradiction. Thus, $L = 0$.

($\mathcal{B}_4$) It is clear that, for all $x, y \in X$ such that $d(x, y) > 0$ and $d(Tx, Ty) > 0$, Theorem 4 guarantees that

$$\varrho_\phi(d(Tx, Ty), d(x, y)) = \phi(d(x, y)) - d(Tx, Ty) > 0.$$

($\mathcal{B}_5$) Let $\{(a_n, b_n)\}$ be a T-sequence such that $\{b_n\} \to 0$ and $\varrho_\phi(a_n, b_n) \geq 0$ for all $n \in \mathbb{N}$. Then, for all $n \in \mathbb{N}$,

$$0 \leq \varrho_\phi(a_n, b_n) = \phi(b_n) - a_n,$$

which means that $0 \leq a_n \leq \phi(b_n) \leq b_n$. Therefore, $\{b_n\} \to 0$ implies $\{a_n\} \to 0$. □

Theorem 7. *Theorem 5 follows from Theorems 2 and 3.*

Proof. From Theorem 6, every Meir–Keeler contraction is an ample spectrum contraction that also verifies ($\mathcal{B}'_2$) and ($\mathcal{B}_5$), thus Theorems 2 and 3 are applicable in order to conclude that every Meir–Keeler contraction has a unique fixed point. □

5.2. Samet et al.'s Contractions

In [9], Samet et al. introduced the following kind of contractions and proved the following results. Let us denote by Ψ the family of nondecreasing functions $\psi : [0, \infty) \to [0, \infty)$ such that $\Sigma_{n \in \mathbb{N}} \psi^n(t) < \infty$ for each $t > 0$, where ψ^n is the nth iterate of ψ.

Definition 7. Let (X,d) be a metric space and $T : X \to X$ be a given mapping. We say that T is an α-ψ-contractive mapping if there exist two functions $\alpha : X \times X \to [0,\infty)$ and $\psi \in \Psi$ such that

$$\alpha(x,y)\, d(Tx, Ty) \leq \psi(d(x,y)) \quad \text{for all } x, y \in X. \tag{9}$$

The main results in [9] can be summarized as follows.

Theorem 8 (Samet, Vetro and Vetro [9], Theorems 2.1, 2.2 and 2.3). *Let (X,d) be a complete metric space and $T : X \to X$ be an α-ψ-contractive mapping satisfying the following conditions:*

(i) T is α-admissible (that is, if $\alpha(x,y) \geq 1$, then $\alpha(Tx, Ty) \geq 1$);
(ii) there exists $x_0 \in X$ such that $\alpha(x_0, Tx_0) \geq 1$; and
(iii) at least, one of the following conditions holds:

(iii.1) T is continuous; or
(iii.2) if $\{x_n\}$ is a sequence in X such that $\alpha(x_n, x_{n+1}) \geq 1$ for all n and $\{x_n\} \to x \in X$ as $n \to \infty$, then $\alpha(x_n, x) \geq 1$ for all n.

Then, T has a fixed point, that is, there exists $x^* \in X$ such that $Tx^* = x^*$.
Furthermore, adding the condition:

(H) for all $x, y \in X$, there exists $z \in X$ such that $\alpha(x,z) \geq 1$ and $\alpha(y,z) \geq 1$,

we obtain uniqueness of the fixed point of T.

To show that the previous theorem can be seen as a consequence of our main results, we present the following statement in which we use a more general class of auxiliary functions.

Theorem 9. *Let (X,d) be a metric space and $T : X \to X$ be a given mapping. Assume that there exist two functions $\alpha : X \times X \to [0,\infty)$ and $\psi : [0,\infty) \to [0,\infty)$ such that ψ is nondecreasing, $\lim_{n \to \infty} \psi^n(t) = 0$ for all $t > 0$, and also*

$$\alpha(x,y)\, d(Tx, Ty) \leq \psi(d(x,y)) \quad \text{for all } x, y \in X. \tag{10}$$

Then, T is an ample spectrum contraction with respect to $\mathcal{S}_\alpha$ that also verifies $(\mathcal{B}'_2)$ and $(\mathcal{B}_5)$.

Proof. Let $\mathcal{S}_\alpha$ be the binary relation on X given in (1). Let $A = \text{ran}(d)$ and let define $\gamma : A \to \mathbb{R}$ and $\varrho_\gamma : A \times A \to \mathbb{R}$ by, for all $t, s \in A$,

$$\gamma(s) = \inf\left(\{\alpha(x,y) : d(x,y) = s\}\right),$$
$$\varrho_\gamma(t,s) = \psi(s) - t\gamma(s).$$

Notice that γ is well defined because if $s \in A = \text{ran}(d)$, then there exist $x_s, y_s \in X$ such that $d(x_s, y_s) = s$, and we can take infimum in a nonempty, subset of non-negative real numbers. Furthermore, as $\gamma(d(x,y)) \leq \alpha(x,y)$ for all $x, y \in X$, then, by (10),

$$\varrho_\gamma(d(Tx, Ty), d(x,y)) = \psi(d(x,y)) - d(Tx, Ty)\gamma(d(x,y))$$
$$\geq \psi(d(x,y)) - d(Tx, Ty)\alpha(x,y) \geq 0.$$

Hence, $(\mathcal{B}_4)$ holds. Let us prove the rest of properties.
$(\mathcal{B}'_2)$ Let $x_1, x_2 \in X$ be two points such that

$$T^n x_1 \mathcal{S}^*_\alpha T^n x_2 \quad \text{and} \quad \varrho_\phi(d\left(T^{n+1}x_1, T^{n+1}x_2\right), d\left(T^n x_1, T^n x_2\right)) \geq 0 \quad \text{for all } n \in \mathbb{N}.$$

Since $T^n x_1 S_\alpha^* T^n x_1$, then $\alpha(T^n x_1, T^n x_2) \geq 1$ and $T^n x_1 \neq T^n x_2$ for all $n \in \mathbb{N}$. By using Equation (10) and Proposition 1, for all $n \in \mathbb{N}$,

$$d(T^{n+1} x_1, T^{n+1} x_2) \leq \alpha(T^n x_1, T^n x_2)\, d(TT^n x_1, TT^n x_2)$$
$$\leq \psi\left(d(T^n x_1, T^n x_2)\right) \leq d(T^n x_1, T^n x_2).$$

As $\{d(T^n x_1, T^n x_2)\}$ is a bounded-below non-increasing sequence of real numbers, it is convergent. Let $L \geq 0$ be its limit. Hence,

$$0 \leq L \leq d(T^{n+1} x_1, T^{n+1} x_2) \leq \psi(d(T^n x_1, T^n x_2)) \leq d(T^n x_1, T^n x_2) \quad \text{for all } n \in \mathbb{N}.$$

As ψ is nondecreasing, for all $n \in \mathbb{N}$,

$$d(T^n x_1, T^n x_2) \leq \psi(d(T^{n-1} x_1, T^{n-1} x_2)) \leq \psi^2(d(T^{n-2} x_1, T^{n-2} x_2)) \leq \ldots \leq \psi^n(d(x_1, x_2)).$$

Taking into account that $d(x_1, x_2) > 0$, then $\lim_{n \to \infty} \psi^n(d(x_1, x_2)) = 0$, and letting $n \to \infty$ in

$$0 \leq L \leq d(T^n x_1, T^n x_2) \leq \psi^n(d(x_1, x_2)),$$

we conclude that $L = \lim_{n \to \infty} d(T^n x_1, T^n x_2) = 0$.

(B_2) It follows from (B_2').

(B_3) Let $\{(a_n, b_n)\} \subseteq A \times A$ be a $(T, \mathcal{S}_\alpha)$-sequence such that $\{a_n\}$ and $\{b_n\}$ converge to the same limit $L \geq 0$ and verifying that $L < a_n$ and $\varrho_\gamma(a_n, b_n) \geq 0$ for all $n \in \mathbb{N}$. By definition, there are two sequences $\{x_n\}, \{y_n\} \subseteq X$ such that

$$x_n S_\alpha y_n, \quad a_n = d(T x_n, T y_n) > 0 \quad \text{and} \quad b_n = d(x_n, y_n) > 0 \quad \text{for all } n \in \mathbb{N}.$$

Hence, $\alpha(x_n, y_n) \geq 1$ for all $n \in \mathbb{N}$. To prove that $L = 0$, we reason by contradiction. Assume that $L > 0$. By Property ($\mathcal{P}_{12}$) of Proposition 1, $\psi(L) < L$. It follows from Equation (10) that

$$\psi(L) < L < a_n = d(T x_n, T y_n) \leq \alpha(x_n, y_n)\, d(T x_n, T y_n) \leq \psi(d(x_n, y_n)) \leq d(x_n, y_n) = b_n. \quad (11)$$

Since $\{b_n\} \to L$, then $\lim_{n \to \infty} \psi(d(x_n, y_n)) = L$. As ψ is nondecreasing, the following limit exists and takes the value

$$\lim_{s \to L^+} \psi(s) = \lim_{n \to \infty} \psi(d(x_n, y_n)) = L.$$

As ψ is nondecreasing, $\psi(L) \leq \psi(s) \leq \psi(t)$ for all $L \leq s \leq t$, so

$$\psi(L) < L = \lim_{s \to L^+} \psi(s) \leq \psi(t) \quad \text{for all } t \in (L, \infty).$$

Taking in mind that $L \leq \psi(t)$ for all $t \in (L, \infty)$, next, we distinguish two cases.

(Case 1) Assume that $\psi(t) > L$ for all $t \in (L, \infty)$. In this case, let $t_0 \in (L, \infty)$ be arbitrary. Then, $\psi(t_0) > L$. Therefore, $\psi^2(t_0) = \psi(\psi(t_0)) > L$. Repeating this argument, $\psi^3(t_0) = \psi(\psi^2(t_0)) > L$. Similarly, by induction, $\psi^n(t_0) > L$ for all $n \in \mathbb{N}$, which contradicts the fact that $\lim_{n \to \infty} \psi^n(t_0) = 0$.

(Case 2) Assume that there exists $L' > L$ such that $\psi(L') = L$. In this case, as ψ is nondecreasing, for all $t \in (L, L']$, we have that $L \leq \psi(t) \leq \psi(L') = L$, so $\psi(t) = L$ for all $t \in (L, L']$. Since $\{b_n = d(x_n, y_n)\} \searrow L^+$, there exists $n_0 \in \mathbb{N}$ such that $d(x_{n_0}, y_{n_0}) \in (L, L']$. Hence, $\psi(d(x_{n_0}, y_{n_0})) = L$, which contradicts the strict inequality in Equation (11) because

$$L < a_{n_0} \leq \psi(d(x_{n_0}, y_{n_0})).$$

In any case, we get a contradiction, so $L = 0$.

($\mathcal{B}_5$) Let $\{(a_n, b_n)\}$ be a $(T, \mathcal{S}_\alpha)$-sequence such that $\{b_n\} \to 0$ and $\varrho_\gamma(a_n, b_n) \geq 0$ for all $n \in \mathbb{N}$. By definition, there exist two sequences $\{x_n\}, \{y_n\} \subseteq X$ such that

$$x_n \mathcal{S}_\alpha y_n, \quad a_n = d(Tx_n, Ty_n) > 0 \quad \text{and} \quad b_n = d(x_n, y_n) > 0 \quad \text{for all } n \in \mathbb{N}.$$

In particular, $\alpha(x_n, y_n) \geq 1$ for all $n \in \mathbb{N}$. It follows from Equation (10) that

$$0 < a_n = d(Tx_n, Ty_n) \leq \alpha(x_n, y_n) \, d(Tx_n, Ty_n) \leq \psi(d(x_n, y_n)) \leq d(x_n, y_n) = b_n.$$

Since $\{b_n\} \to 0$, then $\{a_n\} \to 0$. □

Corollary 1. *Every Samet et al.'s α-ψ-contraction (in the sense of Definition 7) is an ample spectrum contraction with respect to $\mathcal{S}_\alpha$ that also verifies $(\mathcal{B}'_2)$ and $(\mathcal{B}_5)$.*

Proof. It follows from the fact that, if $\psi \in \Psi$, then Theorem 9 is applicable because ψ is nondecreasing and $\lim_{n \to \infty} \psi^n(t) = 0$ for all $t > 0$ (recall Proposition 1). □

Theorem 10. *Theorem 8 immediately follows from Theorems 2 and 3.*

Proof. By Corollary 1, every Samet et al.'s α-ψ-contraction is an ample spectrum contraction with respect to $\mathcal{S}_\alpha$ that also verifies $(\mathcal{B}'_2)$ and $(\mathcal{B}_5)$, thus Theorems 2 and 3 are applicable. □

5.3. Some Meditations about a Nonsymmetric Condition

In [1], Khojasteh et al. introduced the notion of simulation function as a mapping $\zeta : [0, \infty) \times [0, \infty) \to \mathbb{R}$ satisfying the following conditions:

(ζ_1) $\zeta(0, 0) = 0$;
(ζ_2) $\zeta(t, s) < s - t$ for all $t, s > 0$; and
(ζ_3) if $\{t_n\}, \{s_n\}$ are sequences in $(0, \infty)$ such that $\lim_{n \to \infty} t_n = \lim_{n \to \infty} s_n > 0$, then

$$\limsup_{n \to \infty} \zeta(t_n, s_n) < 0.$$

Shortly after, Roldán López de Hierro et al. [2] pointed out that Condition (ζ_3) is symmetric in both arguments of ζ, which is not necessary. Hence, these authors introduced the following variation in Axiom (ζ_3):

(ζ_3') if $\{t_n\}, \{s_n\}$ are sequences in $(0, \infty)$ such that $\lim_{n \to \infty} t_n = \lim_{n \to \infty} s_n > 0$ and $t_n < s_n$ for all $n \in \mathbb{N}$, then

$$\limsup_{n \to \infty} \zeta(t_n, s_n) < 0.$$

In this way, they removed the symmetry of a key function involved in the contractivity condition. After that, Roldán López de Hierro and Shahzad [3] presented the concept of $\mathcal{R}$-contraction, which is intimately associated to an $\mathcal{R}$-function $\varrho : A \times A \to \mathbb{R}$. Such kind of functions must satisfy the following conditions (see [3], Definition 12):

(ϱ_1) If $\{a_n\} \subset (0, \infty) \cap A$ is a sequence such that $\varrho(a_{n+1}, a_n) > 0$ for all $n \in \mathbb{N}$, then $\{a_n\} \to 0$.
(ϱ_2) If $\{a_n\}, \{b_n\} \subset (0, \infty) \cap A$ are two sequences converging to the same limit $L \geq 0$ and verifying that $L < a_n$ and $\varrho(a_n, b_n) > 0$ for all $n \in \mathbb{N}$, then $L = 0$.

Questions immediately arise: Why did the authors impose

$$L < a_n \quad \text{for all } n \in \mathbb{N} \tag{12}$$

in Assumption (ϱ_2)? Why did they not consider

$$L < b_n \quad \text{for all } n \in \mathbb{N} \tag{13}$$

rather than Equation (12)? A first response we can give is that both assumptions are interesting in order to remove the symmetry in the variables of ϱ in Assumption (ϱ_2) because the role of the sequence $\{a_n\}$ is different from the role of $\{b_n\}$. However, are Equations (12) and (13) equivalent? The response is no: we do believe that the condition in Equation (12) is better than the one in Equation (13). We justify it by the following fact: using the hypothesis in Equation (12), it is easy to check that every Meir–Keeler condition is an R-condition (see Theorem 25 in [3]). However, if we have only assumed that Equation (13) holds, then some Meir–Keeler contractions would not have been R-contractions. To illustrate it, we modify Example 2 in the following way.

Example 3. *Let $X = [0,1] \cup C \cup D$, where $C = \{10m \in \mathbb{N} : m \in \mathbb{N}^*\}$ and $D = \{10m + 1 + \frac{1}{m} \in \mathbb{N} : m \in \mathbb{N}^*\}$. If X is furnished with the Euclidean metric $d_E(x,y) = |x-y|$ for all $x,y \in X$, then (X, d_E) is a complete metric space. Let $T : X \to X$ be the self-mapping defined by*

$$Tx = \begin{cases} \dfrac{x}{4}, & \text{if } x \in [0,1], \\ 0, & \text{if } x = 10m \in C \quad \text{(for some } m \in \mathbb{N}^*\text{)}, \\ 1 - \dfrac{1}{2m}, & \text{if } x = 10m + 1 + \frac{1}{m} \in D \quad \text{(for some } m \in \mathbb{N}^*\text{);} \end{cases}$$

Notice that $Tx \in [0,1)$ for all $x \in X$. Therefore,

$$d_E(Tx, Ty) < 1 \quad \text{for all } x,y \in X. \tag{14}$$

Let us show that T is a Meir–Keeler contraction in (X, d_E). Indeed, let $\phi : [0, \infty) \to [0, \infty)$ be the function given by

$$\phi(t) = \begin{cases} \dfrac{t}{2}, & \text{if } t \in [0,1], \\ 1, & \text{if } t > 1. \end{cases}$$

Clearly, ϕ is an L-function, and we claim that Equation (8) holds. Let $x, y \in X$ be such that $d(x,y) > 0$. Suppose, without loss of generality, that $x < y$.

- If $x, y \in [0,1]$, then $d_E(x,y) \leq 1$ and

$$d_E(Tx, Ty) = d_E\left(\frac{x}{4}, \frac{y}{4}\right) = \left|\frac{x}{4} - \frac{y}{4}\right| = \frac{|x-y|}{4} < \frac{|x-y|}{2} = \phi(d_E(x,y)).$$

- If $x \in [0,1]$ and $y \in C \cup D$, then $d_E(x,y) > 1$, and it follows from Equation (14) that

$$d_E(Tx, Ty) < 1 = \phi(d_E(x,y)).$$

- If $x, y \in C \cup D$, then $d_E(x,y) > 1$ and, similarly, $d_E(Tx, Ty) < 1 = \phi(d_E(x,y))$.

In any case, Equation (8) holds and Theorem 4 ensures us that T is a Meir–Keeler contraction in (X, d_E). In fact, Theorem 21 in [3] guarantees that the function $\varrho_\phi : [0, \infty) \times [0, \infty) \to \mathbb{R}$ given by

$$\varrho_\phi(t,s) = \phi(s) - t \quad \text{for all } t, s \in [0, \infty),$$

is an R-function on $[0, \infty)$ verifying (ϱ_3). In particular, it satisfies Axiom (ϱ_2). Let us show that ϱ_ϕ would not satisfy (ϱ_2) if we replace Equation (12) with Equation (13). Indeed, let $\{x_n\}_{n \in \mathbb{N}^*}$ and $\{y_n\}_{n \in \mathbb{N}^*}$ be the sequences in X given by

$$x_n = 10n \quad \text{and} \quad y_n = 10n + 1 + \frac{1}{n} \quad \text{for all } n \in \mathbb{N}.$$

Therefore, for all $n \in \mathbb{N}$,

$$a_n = d_E(Tx_n, Ty_n) = d_E\left(0, 1 - \frac{1}{2n}\right) = 1 - \frac{1}{2n} > 0 \quad \text{and}$$

$$b_n = d_E(x_n, y_n) = d_E\left(10n, 10n + 1 + \frac{1}{n}\right) = 1 + \frac{1}{n} > 1.$$

Hence, for all $n \in \mathbb{N}$,

$$\varrho_\phi(a_n, b_n) = \varrho_\phi\left(1 - \frac{1}{2n}, 1 + \frac{1}{n}\right) = \phi\left(1 + \frac{1}{n}\right) - \left(1 - \frac{1}{2n}\right)$$

$$= 1 - \left(1 - \frac{1}{2n}\right) = \frac{1}{2n} > 0$$

However, $L = 1$ is not zero. Therefore, ϱ_ϕ does not satisfy (ϱ_2) if we replace Equation (12) with Equation (13). Thus, in this case, there would be Meir–Keeler contractions that are not R-contractions.

As it can be easily checked, Property (ϱ_2) that R-functions must satisfy leads to Condition $(\mathcal{A}_3)$ for $(\mathcal{A}, \mathcal{S})$-contractions and Condition $(\mathcal{B}_3)$ for ample spectrum contractions.

$(\mathcal{B}_3)$ If $\{(a_n, b_n)\} \subseteq A \times A$ is a $(T, \mathcal{S}^*)$-sequence such that $\{a_n\}$ and $\{b_n\}$ converge to the same limit $L \geq 0$ and verifying that $L < a_n$ and $\varrho(a_n, b_n) \geq 0$ for all $n \in \mathbb{N}$, then $L = 0$.

If we have assumed the condition in Equation (13) rather than the condition in Equation (12) in $(\mathcal{B}_3)$, then the same arguments given in Example 3 prove that there would be Meir–Keeler contractions that are not ample spectrum contractions. As a consequence, we conclude that the assumption in Equation (12) is more appropriate than the one in Equation (13) in the context of fixed point theory.

Nevertheless, in the next subsection, we are going to show that, under some very recent contractivity conditions, they would be equivalent.

5.4. Shahzad et al.'s Contractions

In [10], Shahzad et al. presented some coincidence point results for a new class of contractive mappings that they called (α, ψ, ϕ)-contractions. They used the following kind of auxiliary functions.

Definition 8 (Roldán López de Hierro [10], Definition 3.5). *Let $\mathcal{F}_A$ be the family of all pairs (ψ, ϕ) where $\psi, \phi : [0, \infty) \to [0, \infty)$ are two functions verifying the following two conditions:*

$(\mathcal{F}_A^1)$ *If $\{a_n\} \subset (0, \infty)$ is a sequence such that $\psi(a_{n+1}) \leq \phi(a_n)$ for all $n \in \mathbb{N}$, then $\{a_n\} \to 0$.*
$(\mathcal{F}_A^2)$ *If $\{a_n\}, \{b_n\} \subset [0, \infty)$ are two sequences converging to the same limit L and such that $L < a_n$ and $\psi(b_n) \leq \phi(a_n)$ for all $n \in \mathbb{N}$, then $L = 0$.*

As a consequence of their main coincidence results, they presented the following statement (see the necessary preliminaries in [10]).

Theorem 11 (Shahzad, Karapınar and Roldán López de Hierro [10], Theorem 6.1). *Let (X, d) be a metric space, let $\alpha : X \times X \to [0, \infty)$ be a function and let $T : X \to X$ be a mapping such that the following conditions are fulfilled:*

1. *there exists a subset $A \subseteq X$ such that $T(X) \subseteq A$ and (A, d) is complete;*
2. *α is transitive and T is α-admissible;*
3. *there exists $(\psi, \phi) \in \mathcal{F}_A$ such that*

$$\alpha(x, y) \, \psi\left(d(Tx, Ty)\right) \leq \phi\left(d(x, y)\right) \quad \text{for all } x, y \in X; \tag{15}$$

and

4. at least one of the following conditions holds:

    (a) there exists $x_0 \in X$ such that $\alpha(x_0, Tx_0) \geq 1$ and T is (d, α)-right-continuous; or
    (b) there exists $x_0 \in X$ such that $\alpha(Tx_0, x_0) \geq 1$ and T is (d, α)-left-continuous.

Then, T has, at least, a fixed point.
Additionally, assume that $\phi(0) = 0$, $\psi^{-1}(\{0\}) = \{0\}$, and the following property holds:

($\mathcal{U}$) for all fixed points x and y of T, there exists $z \in X$ such that z is, at the same time, α-comparable to x and to y.

Then, T has a unique fixed point.

In the following definition, we modify the second condition.

Definition 9. Let $\mathcal{G}_A$ be the family of all pairs (ψ, ϕ) where $\psi, \phi : [0, \infty) \to [0, \infty)$ are two functions verifying the following two conditions:

($\mathcal{F}_A^1$) If $\{a_n\} \subset (0, \infty)$ is a sequence such that $\psi(a_{n+1}) \leq \phi(a_n)$ for all $n \in \mathbb{N}$, then $\{a_n\} \to 0$.
($\mathcal{G}_A^2$) If $\{a_n\}, \{b_n\} \subset [0, \infty)$ are two sequences converging to the same limit L and such that $L < b_n$ and $\psi(b_n) \leq \phi(a_n)$ for all $n \in \mathbb{N}$, then $L = 0$.

The same theorem can be proved in this case.

Theorem 12. Let (X, d) be a metric space, let $\alpha : X \times X \to [0, \infty)$ be a function and let $T : X \to X$ be a mapping such that the following conditions are fulfilled:

1. There exists a subset $A \subseteq X$ such that $T(X) \subseteq A$ and (A, d) is complete.
2. α is transitive and T is α-admissible.
3. There exists $(\psi, \phi) \in \mathcal{G}_A$ such that

$$\alpha(x, y) \psi(d(Tx, Ty)) \leq \phi(d(x, y)) \quad \text{for all } x, y \in X. \tag{16}$$

4. At least one of the following conditions holds:

    (a) there exists $x_0 \in X$ such that $\alpha(x_0, Tx_0) \geq 1$ and T is (d, α)-right-continuous; or
    (b) there exists $x_0 \in X$ such that $\alpha(Tx_0, x_0) \geq 1$ and T is (d, α)-left-continuous.

Then, T has, at least, a fixed point.
Additionally, assume that $\phi(0) = 0$, $\psi^{-1}(\{0\}) = \{0\}$, and the following property holds:

($\mathcal{U}$) For all fixed points x and y of T, there exists $z \in X$ such that z is, at the same time, α-comparable to x and to y.

Then, T has a unique fixed point.

Let us show how this last result can be deduced from Theorems 2 and 3. The key is the following result.

Lemma 2. Let (X, d) be a metric space, let $\alpha : X \times X \to [0, \infty)$ be a function and let $T : X \to X$ be a mapping such that the following conditions are fulfilled:

1. There exists $(\psi, \phi) \in \mathcal{G}_A$ such that

$$\alpha(x, y) \psi(d(Tx, Ty)) \leq \phi(d(x, y)) \quad \text{for all } x, y \in X. \tag{17}$$

2. There exist two distinct points $x_0, x_1 \in X$ such that $\alpha(x_0, x_1) \geq 1$.

Then, T is an ample spectrum contraction with respect to a function ϱ and $\mathcal{S}_\alpha$ that also verifies $(\mathcal{B}'_2)$.

Proof. Let us consider

$$A = \{ d(x,y) \in [0, \infty) : x, y \in X, \, x\mathcal{S}^*_\alpha y \}$$
$$= \{ d(x,y) \in [0, \infty) : x, y \in X, \, x \neq y, \, \alpha(x,y) \geq 1 \}.$$

As $d(x_0, x_1) \in A$, then A is nonempty. Let us define the function $\gamma : A \to \mathbb{R}$, for all $t \in A$, by

$$\gamma(t) = \inf (\{ \alpha(x,y) : x, y \in X, \, x\mathcal{S}^*_\alpha y \text{ and } d(x,y) = t \}).$$

To prove that γ is well defined, let $t \in A$ be arbitrary and let

$$\Omega_t = \{ \alpha(x,y) : x, y \in X, \, x\mathcal{S}^*_\alpha y \text{ and } d(x,y) = t \}.$$

By definition, as $t \in A$, there exist $x_t, y_t \in X$ such that $x_t \mathcal{S}^*_\alpha y_t$ and $t = d(x_t, y_t)$. Therefore, $\alpha(x_t, y_t) \in \Omega_t$, so this set is nonempty. Moreover, let $x, y \in X$ be arbitrary points such that $x\mathcal{S}^*_\alpha y$ and $d(x,y) = t$. Hence, $\alpha(x,y) \geq 1$. This proves that $\alpha(x,y) \geq 1$ for all number $\alpha(x,y) \in \Omega_t$. Taking into account that Ω_t is nonempty and bounded below by 1, we can take infimum, which means that $\gamma(t)$ is well defined. In particular, we have proved the following facts:

$$\gamma(t) = \inf \Omega_t \geq 1 \quad \text{for all } t \in A; \tag{18}$$
$$\gamma(d(x,y)) \leq \alpha(x,y) \quad \text{for all } x, y \in X \text{ such that } x\mathcal{S}^*_\alpha y. \tag{19}$$

Considering the pair $(\psi, \phi) \in \mathcal{G}_A$, let $\varrho : A \times A \to \mathbb{R}$ be defined, for all $t, s \in A$, by

$$\varrho(t,s) = \phi(s) - \gamma(s) \, \psi(t) \quad \text{for all } t, s \in A.$$

We claim that T is an ample spectrum contraction with respect to ϱ and $\mathcal{S}_\alpha$ that also verifies $(\mathcal{B}'_2)$. We demonstrate each condition. $(\mathcal{B}_1)$ is obvious.

$(\mathcal{B}_4)$ Let $x, y \in X$ be arbitrary points such that $x\mathcal{S}^*_\alpha y$ and $Tx\mathcal{S}^*_\alpha Ty$, that is, $\alpha(x,y) \geq 1$, $\alpha(Tx, Ty) \geq 1$, $x \neq y$ and $Tx \neq Ty$. Therefore, applying Equation (17),

$$\alpha(x,y) \, \psi(d(Tx, Ty)) \leq \phi(d(x,y)). \tag{20}$$

In particular, it follows from Equations (19) and (20) that

$$\varrho(d(Tx, Ty), d(x,y)) = \phi(d(x,y)) - \gamma(d(x,y)) \, \psi(d(Tx, Ty))$$
$$\geq \phi(d(x,y)) - \alpha(x,y) \, \psi(d(Tx, Ty)) \geq 0,$$

so $(\mathcal{B}_4)$ holds.

$(\mathcal{B}'_2)$ Let $x_1, x_2 \in X$ be two points such that

$$T^n x_1 \mathcal{S}^*_\alpha T^n x_2 \quad \text{and} \quad \varrho(d(T^{n+1}x_1, T^{n+1}x_2), d(T^n x_1, T^n x_2)) \geq 0 \quad \text{for all } n \in \mathbb{N}.$$

Notice that $T^n x_1 \mathcal{S}^*_\alpha T^n x_2$ and $T^{n+1} x_1 \mathcal{S}^*_\alpha T^{n+1} x_2$ imply that $d(T^n x_1, T^n x_2)$ and $d(T^{n+1} x_1, T^{n+1} x_2)$ belong to A. Let

$$a_n = d(T^n x_1, T^n x_2) > 0 \quad \text{for all } n \in \mathbb{N}.$$

In particular, as $\gamma \geq 1$, then

$$0 \leq \varrho(d\left(T^{n+1}x_1, T^{n+1}x_2\right), d\left(T^n x_1, T^n x_2\right)) = \varrho\left(a_{n+1}, a_n\right)$$
$$= \phi\left(a_n\right) - \gamma\left(a_n\right)\psi\left(a_{n+1}\right) \leq \phi\left(a_n\right) - \psi\left(a_{n+1}\right),$$

that is, $\psi\left(a_{n+1}\right) \leq \phi\left(a_n\right)$, for all $n \in \mathbb{N}$. Since $(\phi, \psi) \in \mathcal{G}_\mathcal{A}$, Condition $(\mathcal{F}_\mathcal{A}^1)$ implies that $\{a_n\} \to 0$, that is, $\{d\left(T^n x_1, T^n x_2\right)\} \to 0$, which means that $(\mathcal{B}_2')$ holds.

$(\mathcal{B}_2)$ It immediately follows from $(\mathcal{B}_2')$.

$(\mathcal{B}_3)$ Let $\{(a_n', b_n')\} \subseteq A \times A$ be a $(T, \mathcal{S}_\alpha^*)$-sequence such that $\{a_n'\}$ and $\{b_n'\}$ converge to the same limit $L \geq 0$ and verifying that $L < a_n'$ and $\varrho(a_n', b_n') \geq 0$ for all $n \in \mathbb{N}$. By definition, there exist two sequences $\{x_n\}, \{y_n\} \subseteq X$ such that

$$x_n \mathcal{S}^* y_n, \quad Tx_n \mathcal{S}^* Ty_n, \quad a_n' = d(Tx_n, Ty_n) > 0 \quad \text{and} \quad b_n' = d(x_n, y_n) > 0 \quad \text{for all } n \in \mathbb{N}.$$

As $\gamma \geq 1$, then

$$0 \leq \varrho(a_n', b_n') = \phi\left(b_n'\right) - \gamma\left(b_n'\right)\psi\left(a_n'\right) \leq \phi\left(b_n'\right) - \psi\left(a_n'\right),$$

that is, $\psi\left(a_n'\right) \leq \phi\left(b_n'\right)$, for all $n \in \mathbb{N}$. Since $(\phi, \psi) \in \mathcal{G}_\mathcal{A}$, Condition $(\mathcal{G}_\mathcal{A}^2)$ (applied to $\{a_n\} = \{b_n'\}$ and $\{b_n\} = \{a_n'\}$) implies that $L = 0$, which means that $(\mathcal{B}_3)$ holds.

As a consequence, we conclude that T is an ample spectrum contraction with respect to ϱ and $\mathcal{S}_\alpha$ that also verifies $(\mathcal{B}_2')$. □

Lemma 2 permits us to show that Theorem 12 is a particular case of the above-presented main statements.

Theorem 13. *Theorem 12 follows from Theorems 2 and 3.*

Proof. Assume that all the hypotheses of Theorem 12 hold. For instance, assume that there exists $x_0 \in X$ such that $\alpha(x_0, Tx_0) \geq 1$ and T is (d, α)-right-continuous (notice that Condition (4.b) requires a version of Theorems 2 and 3 in which T is non-increasing). Let $\{x_{n+1} = Tx_n\}_{n \geq 0}$ be the Picard sequence of T based on x_0. If there exists some $n_0 \in \mathbb{N}$ such that $x_{n_0+1} = x_{n_0}$, then x_{n_0} is a fixed point of T, and $\{x_n\}$ converges to such point. In this case, the part about existence of a fixed point of T is finished. On the contrary case, assume that $x_n \neq x_{n+1}$ for all $n \in \mathbb{N}$. Let $\mathcal{S}_\alpha$ be the binary relation on X given, for $x, y \in X$, by

$$x \mathcal{S}_\alpha y \quad \text{if} \quad \alpha(x, y) \geq 1. \tag{21}$$

By Lemma 1:

- As α is transitive, $\mathcal{S}_\alpha$ is transitive.
- As T is α-admissible, T is $\mathcal{S}_\alpha$-nondecreasing.
- As T is (d, α)-right-continuous, T is $\mathcal{S}_\alpha$-nonincreasing-continuous, thus T is $\mathcal{S}_\alpha$-strictly-increasing-continuous (T satisfies Item (a) of Theorem 2).

By Hypothesis 1 of Theorem 12, there exists a subset $A \subseteq X$ such that $T(X) \subseteq A$ and (A, d) is complete. In particular, $T(X)$ is $(\mathcal{S}_\alpha, d)$-strictly-increasing-precomplete. Finally, Lemma 2 guarantees that T is a an ample spectrum contraction with respect to ϱ and $\mathcal{S}_\alpha$ that also verifies $(\mathcal{B}_2')$. As all hypotheses of Theorem 2 are satisfied, T has at least a fixed point.

Following the statement of Theorem 12, additionally, assume that $\phi(0) = 0$, $\psi^{-1}(\{0\}) = \{0\}$, and the following property holds:

(U) For all fixed points x and y of T, there exists $z \in X$ such that z is, at the same time, α-comparable to x and to y.

Then, Theorem 3 is applicable, thus T has a unique fixed point. □

Remark 3. *Notice that, in fact, we have proved that every Shahzad et al.'s contraction in the sense of Theorem 11 is an ample spectrum contraction with respect to an appropriate function ϱ.*

5.5. Wardowski's F-Contractions

Definition 10 (Wardowski [11], Definition 2.1). *Given a function $F : (0, \infty) \to \mathbb{R}$, let consider the following properties:*

(F_1) *F is strictly increasing, that is, $F(t) < F(s)$ for all $t, s \in (0, \infty)$ such that $t < s$.*
(F_2) *For each sequence $\{t_n\}_{n \in \mathbb{N}}$ of positive real numbers we have that $\{t_n\} \to 0$ if, and only if, $\{F(t_n)\} \to -\infty$.*
(F_3) *There exists $\lambda \in (0,1)$ such that $\lim_{t \to 0^+} t^\lambda F(t) = 0$.*

If (X,d) is a metric space, a mapping $T : X \to X$ is an F-contraction if there exist a positive number $\tau > 0$ and a function $F : (0, \infty) \to \mathbb{R}$ satisfying properties (F_1)-(F_3) such that

$$\tau + F(d(Tx, Ty)) \leq F(d(x,y)) \quad \text{for all } x, y \in X \text{ such that } d(Tx, Ty) > 0.$$

Theorem 14 (Wardowski [11], Theorem 2.1). *Let (X, d) be a complete metric space and let $T : X \to X$ be an F-contraction. Then, T has a unique fixed point $x^* \in X$, and for every $x_0 \in X$ a sequence $\{T^n x_0\}_{n \in \mathbb{N}}$ is convergent to x^*.*

Lemma 3. *Every F-contraction is an ample spectrum contraction.*

Notice that in the following proof we do not use Property (F_3).

Proof. Let (X, d) be a metric space and let $T : X \to X$ be an F-contraction with respect to a constant $\tau > 0$ and a function $F : (0, \infty) \to \mathbb{R}$. Let $\lambda = e^{-\tau} \in (0, 1)$, let $A = [0, \infty)$ and let $\phi : (0, \infty) \to (0, \infty)$ and $\varrho : A \times A \to \mathbb{R}$ be the functions:

$$\phi(t) = \begin{cases} e^{F(t)}, & \text{if } t > 0, \\ 0, & \text{if } t = 0; \end{cases}$$

$$\varrho(t, s) = \lambda \phi(s) - \phi(t) \quad \text{for all } t, s \in [0, \infty)$$

Property (F_1) implies that ϕ is strictly increasing on $(0, \infty)$ and Property (F_2) guarantees that for each sequence $\{t_n\}_{n \in \mathbb{N}}$ of positive real numbers we have that

$$\{t_n\} \to 0 \text{ if, and only if, } \{\phi(t_n)\} \to 0. \tag{22}$$

We claim that T is an ample spectrum contraction with respect to ϱ and the trivial preorder $\mathcal{S}_X$. Property ($\mathcal{B}_1$) is obvious.

($\mathcal{B}_2$) Let $\{x_n\} \subseteq X$ be a Picard sequence of T such that

$$x_n \neq x_{n+1} \quad \text{and} \quad \varrho(d(x_{n+1}, x_{n+2}), d(x_n, x_{n+1})) \geq 0 \quad \text{for all } n \in \mathbb{N}.$$

Therefore, for all $n \in \mathbb{N}$, $d(x_n, x_{n+1}) > 0$ and

$$0 \leq \varrho(d(x_{n+1}, x_{n+2}), d(x_n, x_{n+1})) = \lambda \phi(d(x_n, x_{n+1})) - \phi(d(x_{n+1}, x_{n+2})),$$

so

$$0 \leq \phi(d(x_{n+1}, x_{n+2})) \leq \lambda \phi(d(x_n, x_{n+1})).$$

In particular, $\{\phi(d(x_n, x_{n+1}))\} \to 0$, and the property in Equation (22) guarantees that $\{d(x_n, x_{n+1})\} \to 0$.

($\mathcal{B}_3$) Let $\{(a_n, b_n)\} \subseteq A \times A$ be a $(T, \mathcal{S}_X^*)$-sequence such that $\{a_n\}$ and $\{b_n\}$ converge to the same limit $L \geq 0$ and verifying that $L < a_n$ and $\varrho(a_n, b_n) \geq 0$ for all $n \in \mathbb{N}$. By Definition 3, $a_n > 0$ and $b_n > 0$ for all $n \in \mathbb{N}$. To prove that $L = 0$, assume, by contradiction, that $L > 0$. Notice that for all $n \in \mathbb{N}$,

$$0 \leq \varrho(a_n, b_n) = \lambda \phi(b_n) - \phi(a_n).$$

As ϕ is strictly increasing,

$$0 < \phi(L) < \phi(a_n) \leq \lambda \phi(b_n) < \phi(b_n).$$

This means that $L < a_n < b_n$. Since ϕ is strictly increasing, the following limit exists:

$$L' = \lim_{s \to L^+} \phi(s).$$

Furthermore, $0 < \phi(L) \leq L'$. As $\{a_n\} \to L$, $\{b_n\} \to L$ and $L < a_n < b_n$ for all $n \in \mathbb{N}$, then

$$L' = \lim_{s \to L^+} \phi(s) = \lim_{n \to \infty} \phi(a_n) = \lim_{n \to \infty} \phi(b_n).$$

Taking limit as $n \to \infty$ in $\phi(a_n) \leq \lambda \phi(b_n)$, we deduce that $L' \leq \lambda L'$, which contradicts the fact that $L' > 0$. Therefore, $L = 0$.

($\mathcal{B}_4$) Let $x, y \in X$ be two points such that $Tx \neq Ty$. In particular, $d(Tx, Ty) > 0$. Hence,

$$\tau + F(d(Tx, Ty)) \leq F(d(x,y)) \quad \Leftrightarrow \quad e^{\tau + F(d(Tx,Ty))} \leq e^{F(d(x,y))}$$
$$\Leftrightarrow \quad e^{F(d(Tx,Ty))} \leq e^{-\tau} e^{F(d(x,y))} \quad \Leftrightarrow \quad \phi(d(Tx, Ty)) \leq \lambda \phi(d(x,y))$$
$$\Leftrightarrow \quad \lambda \phi(d(x,y)) - \phi(d(Tx, Ty)) \geq 0 \quad \Leftrightarrow \quad \varrho(d(Tx, Ty), d(x,y)) \geq 0.$$

Therefore, T is an ample spectrum contraction with respect to ϱ and $\mathcal{S}_X$. □

As a consequence, Theorem 14 is a simple consequence of Theorems 2 and 3.

Finally, we point out that the present techniques can be easily generalized to guarantee existence and uniqueness of multidimensional coincidence/fixed points following the techniques described in [19–25].

Author Contributions: Conceptualization, A.F.R.L.d.H. and N.S.; Methodology, A.F.R.L.d.H. and N.S.; Writing-Original Draft Preparation, A.F.R.L.d.H. and N.S.; Writing-Review & Editing, A.F.R.L.d.H. and N.S.

Funding: This article was funded by the Deanship of Scientific Research (DSR), King Abdulaziz University, Jeddah.

Acknowledgments: The authors acknowledge with thanks DSR for financial support. A.F. Roldán López de Hierro is grateful to Junta de Andalucía,y project FQM-268 of the Andalusian CICYE and Project TIN2017-89517-P of the Ministerio de Economía, Industria y Competitividad.

Conflicts of Interest: The authors declare no conflict of interest.

References

1. Khojasteh, F.; Shukla, S.; Radenović, S. A new approach to the study of fixed point theory for simulation functions. *Filomat* **2015**, *29*, 1189–1194. [CrossRef]
2. Roldán López de Hierro, A.F.; Karapınar, E.; Roldán López de Hierro, C.; Martínez-Moreno, J. Coincidence point theorems on metric spaces via simulation functions. *J. Comput. Appl. Math.* **2015**, *275*, 345–355. [CrossRef]
3. Roldán López de Hierro, A.F.; Shahzad, N. New fixed point theorem under R-contractions. *Fixed Point Theory Appl.* **2015**, *2015*, 345. [CrossRef]

4. Roldán López de Hierro, A.F.; Shahzad, N. Common fixed point theorems under (R,S)-contractivity conditions. *Fixed Point Theory Appl.* **2016**, *2016*, 55. [CrossRef]
5. Shahzad, N.; Roldán López de Hierro, A.F.; Khojasteh, F. Some new fixed point theorems under (A,S)-contractivity conditions. *RACSAM Rev. R. Acad. A* **2017**, *111*, 307–324.
6. Alam, A.; Imdad, M. Relation-theoretic metrical coincidence theorems. *Filomat* **2017**, *31*, 4421–4439. [CrossRef]
7. Perveen, A.; Uddin, I.; Imdad, M. Generalized contraction principle under relatively weaker contraction in partial metric spaces. *Adv. Differ. Equ.* **2019**, *2019*, 88. [CrossRef]
8. Hussain, A.; Kanwal, T.; Mitrović, Z.D.; Radenović, S. Optimal solutions and applications to nonlinear matrix and integral equations via simulation function. *Filomat* **2018**, *32*, 6087–6106. [CrossRef]
9. Samet, B.; Vetro, C.; Vetro, P. Fixed point theorem for α-ψ-contractive type mappings. *Nonlinear Anal.* **2012**, *75*, 2154–2165. [CrossRef]
10. Shahzad, N.; Karapınar, E.; Roldán López de Hierro, A.F. On some fixed point theorems under (α,ψ,ϕ)-contractivity conditions in metric spaces endowed with transitive binary relations. *Fixed Point Theory Appl.* **2015**, *2015*, 124. [CrossRef]
11. Wardowski, D. Fixed points of a new type of contractive mappings in complete metric spaces. *Fixed Point Theory Appl.* **2012**, *2012*, 94. [CrossRef]
12. Agarwal, R.P.; Karapınar, E.; O'Regan, D.; Roldán López de Hierro, A.F. *Fixed Point Theory in Metric Type Spaces*; Springer International Publishing: Cham, Switzerland, 2015.
13. Rus, I.A. *Generalized Contractions and Applications*; Cluj University Press: Cluj-Napoca, Romania, 2001.
14. Lim, T.-C. On characterizations of Meir-Keeler contractive maps. *Nonlinear Anal.* **2001**, *46*, 113–120. [CrossRef]
15. Meir, A.; Keeler, E. A theorem on contraction mappings. *J. Math. Anal. Appl.* **1969**, *28*, 326–329. [CrossRef]
16. Du, W.-S.; Khojasteh, F. New results and generalizations for approximate fixed point property and their applications. *Abstr. Appl. Anal.* **2014**, *2014*, 581267. [CrossRef]
17. Geraghty, M. On contractive mappings. *Proc. Am. Math. Soc.* **1973**, *40*, 604–608. [CrossRef]
18. Chu, S.C.; Diaz, J.B. Remarks on a generalization of Banach's principle of contraction mappings. *J. Math. Anal. Appl.* **1965**, *2*, 440–446. [CrossRef]
19. Al-Mezel, S.A.; Alsulami, H.H.; Karapınar, E.; Roldán, A. Discussion on "Multidimensional coincidence points" via recent publications. *Abstr. Appl. Anal.* **2014**, *2014*, 287492. [CrossRef]
20. Karapınar, E.; Roldán, A. A note on "*n*-tuplet fixed point theorems for contractive type mappings in partially ordered metric spaces". *J. Ineq. Appl.* **2013**, *2013*, 567. [CrossRef]
21. Karapınar, E.; Roldán, A.; Shahzad, N.; Sintunavarat, W. Discussion of coupled and tripled coincidence point theorems for φ-contractive mappings without the mixed g-monotone property. *Fixed Point Theory Appl.* **2014**, *2014*, 92. [CrossRef]
22. Roldán, A.; Martínez-Moreno, J.; Roldán, C.; Karapınar, E. Meir-Keeler type multidimensional fixed point theorems in partially ordered complete metric spaces. *Abstr. Appl. Anal.* **2013**, *2013*, 406026. [CrossRef]
23. Roldán López de Hierro, A.F.; Karapınar, E.; de la Sen, M. Coincidence point theorems in quasi-metric spaces without assuming the mixed monotone property and consequences in G-metric spaces. *Fixed Point Theory Appl.* **2014**, *2014*, 184. [CrossRef]
24. Roldán, A.; Martínez-Moreno, J.; Roldán, C.; Karapınar, E. Some remarks on multidimensional fixed point theorems. *Fixed Point Theory* **2014**, *15*, 545–558.
25. Roldán, A.; Martínez-Moreno, J.; Roldán, C.; Cho, Y.J. Multidimensional fixed point theorems under (ψ,φ)-contractive conditions in partially ordered complete metric spaces. *J. Comput. Appl. Math.* **2015**, *273*, 76–87. [CrossRef]

© 2019 by the authors. Licensee MDPI, Basel, Switzerland. This article is an open access article distributed under the terms and conditions of the Creative Commons Attribution (CC BY) license (http://creativecommons.org/licenses/by/4.0/).

Article

A Characterization of Quasi-Metric Completeness in Terms of α–ψ-Contractive Mappings Having Fixed Points

Salvador Romaguera and Pedro Tirado *

Instituto Universitario de Matemática Pura y Aplicada-IUMPA, Universitat Politècnica de València, 46022 Valencia, Spain; sromague@mat.upv.es
* Correspondence: pedtipe@mat.upv.es

Received: 2 December 2019; Accepted: 17 December 2019; Published: 19 December 2019

Abstract: We obtain a characterization of Hausdorff left K-complete quasi-metric spaces by means of α–ψ-contractive mappings, from which we deduce the somewhat surprising fact that one the main fixed point theorems of Samet, Vetro, and Vetro (see "Fixed point theorems for α–ψ-contractive type mappings", *Nonlinear Anal.* **2012**, *75*, 2154–2165), characterizes the metric completeness.

Keywords: fixed point; quasi-metric space; left K-complete; α–ψ-contractive mapping

1. Introduction and Preliminaries

In their interesting and germinal paper [1], Samet, Vetro, and Vetro obtained various fixed point theorems in terms of α–ψ contractions which allowed them to deduce, in an elegant and direct way, several important and well-known fixed point results from [2–5]. Many authors have continued the research of this type of contractions and their generalizations in different contexts (see e.g., [6–12]). Recently, Fulga and Taş [13] have presented a careful and extensive study for several generalized α–ψ contractions in the realm of quasi-metric spaces.

In this note we obtain a characterization of Hausdorff left K-complete quasi-metric spaces by means of α–ψ-contractive mappings from which we deduce the somewhat surprising fact that one the main fixed point theorems of Samet, Vetro, and Vetro [1] (Theorem 2.2) characterizes the metric completeness (see Corollary 1 at the end of the paper).

Let us recall that the problem of characterizing the metric completeness in term of fixed point theorems has been studied and solved by several authors with different approaches (see e.g., [14–17]) and that this study has been extended in recent years to some types of generalized metric spaces as partial metric spaces [18,19] and quasi-metric spaces [20,21].

In order to help the reader, we recall some notions and properties of quasi-metric spaces which will be used in this paper. Our basic reference is [22].

A quasi-metric space is a pair $(\mathcal{X}, \rho)$ such that $\mathcal{X}$ is a set and ρ is a quasi-metric on $\mathcal{X}$, i.e., ρ is a function from $\mathcal{X} \times \mathcal{X}$ to $[0, \infty)$ such that for all $\zeta, \eta, \theta \in \mathcal{X}$:

(i) $\zeta = \eta$ if and only if $\rho(\zeta, \eta) = \rho(\eta, \zeta) = 0$, and
(ii) $\rho(\zeta, \theta) \leq \rho(\zeta, \eta) + \rho(\eta, \theta)$.

Given a quasi-metric ρ on $\mathcal{X}$ the family $\{B_\rho(\zeta, \varepsilon) : \zeta \in \mathcal{X}, \varepsilon > 0\}$, where $B_\rho(\zeta, \varepsilon) = \{\eta \in \mathcal{X} : \rho(\zeta, \eta) < \varepsilon\}$ for all $\zeta \in \mathcal{X}$ and $\varepsilon > 0$, is a base for a $\mathcal{T}_0$ topology τ_ρ on $\mathcal{X}$.

$(\mathcal{X}, \rho)$ is called a $\mathcal{T}_1$ quasi-metric space if τ_ρ is a $\mathcal{T}_1$ topology, and it is called a Hausdorff quasi-metric space if τ_ρ is a $\mathcal{T}_2$ topology.

A quasi-metric space $(\mathcal{X}, \rho)$ is said to be left K-complete if every left K-Cauchy sequence converges with respect to τ_ρ, where, by a left K-Cauchy sequence we mean a sequence $(\zeta_n)_{n\in\mathbb{N}}$ in $(\mathcal{X}, \rho)$ such that for each $\varepsilon > 0$ there exists $n_\varepsilon \in \mathbb{N}$ satisfying $\rho(\zeta_n, \zeta_m) < \varepsilon$ whenever $n_\varepsilon \leq n \leq m$.

2. Results

We start this section by recalling some known concepts.

As usual, we denote by Ψ the family of nondecreasing functions $\psi : [0, \infty) \to [0, \infty)$ such that $\sum_{n=1}^{\infty} \psi^n(t) < \infty$ for all $t \geq 0$.

Let $\mathcal{X}$ be a set, $\mathcal{T} : \mathcal{X} \to \mathcal{X}$ and $\alpha : \mathcal{X} \times \mathcal{X} \to [0, \infty)$. Following [1] (Definition 2.2), we say that $\mathcal{T}$ is α-admissible if $\alpha(\zeta, \eta) \geq 1$ implies $\alpha(\mathcal{T}\zeta, \mathcal{T}\eta) \geq 1$; $\zeta, \eta \in \mathcal{X}$.

As in the metric case [1] (Definition 2.1), given a quasi-metric space $(\mathcal{X}, \rho)$ we say that a mapping $\mathcal{T} : \mathcal{X} \to \mathcal{X}$ is an α–ψ-contractive mapping if there exist two functions $\alpha : \mathcal{X} \times \mathcal{X} \to [0, \infty)$ and $\psi \in \Psi$ such that $\alpha(\zeta, \eta)\rho(\mathcal{T}\zeta, \mathcal{T}\eta) \leq \psi(\rho(\zeta, \eta))$ for all $\zeta, \eta \in \mathcal{X}$.

The following slight modification of condition (iii) in Theorem 2.2 of [1] constitutes a crucial ingredient in obtaining our main result:

Let $(\mathcal{X}, \rho)$ be a quasi-metric space and $\alpha : \mathcal{X} \times \mathcal{X} \to [0, \infty)$. We say that $(\mathcal{X}, \rho)$ has property (A) (with respect to α) if for any sequence $(\zeta_n)_{n\in\mathbb{N}}$ in $\mathcal{X}$ satisfying $\alpha(\zeta_n, \zeta_{n+1}) \geq 1$ for all $n \in \mathbb{N}$ and such that $\rho(\zeta, \zeta_n) \to 0$ as $n \to \infty$ for some $\zeta \in \mathcal{X}$, it follows that $\alpha(\zeta, \zeta_n) \geq 1$ for all $n \in \mathbb{N}$.

Definition 1. *Given a quasi-metric space $(\mathcal{X}, \rho)$, an α–ψ-contractive mapping $\mathcal{T} : \mathcal{X} \to \mathcal{X}$ will be called an α–ψ-SVV contractive mapping if: (i) $\mathcal{T}$ is α-admissible; (ii) there exists $\zeta_0 \in \mathcal{X}$ such that $\alpha(\zeta_0, \mathcal{T}\zeta_0) \geq 1$; (iii) $(\mathcal{X}, \rho)$ has property (A) (with respect to α).*

By using the preceding definition, Theorem 2.2 of [1] can be reformulated as follows: *Every α–ψ-SVV contractive mapping on a complete metric space has a fixed point.*

Our first result provides a quasi-metric extension of Theorem 2.2 of [1] (its proof is only an adaptation of the original proof of Samet, Vetro, and Vetro).

Theorem 1. *Every α–ψ-SVV contractive mapping on a left K-complete quasi-metric space has a fixed point.*

Proof of Theorem 1. Let $\mathcal{T}$ be an α–ψ-SVV contractive mapping on a Hausdorff left K-complete quasi-metric space $(\mathcal{X}, \rho)$. Then, there exists an α-admissible function such that $\mathcal{T}$ is α–ψ-contractive, $(\mathcal{X}, \rho)$ has property (A), and $\alpha(\zeta_0, \mathcal{T}\zeta_0) \geq 1$ for some $\zeta_0 \in \mathcal{X}$.

For each $n \in \mathbb{N}$ let $\zeta_n := \mathcal{T}^n \zeta_0$. If there exists $m \in \mathbb{N}$ such that $\zeta_{m-1} = \zeta_m$, then ζ_m is a fixed point of $\mathcal{T}$. Assume then that $\zeta_n \neq \zeta_m$ for all $n, m \in \mathbb{N} \cup \{0\}$. Since $\alpha(\zeta_0, \zeta_1) \geq 1$ and $\mathcal{T}$ is α-admissible we deduce that $\alpha(\zeta_n, \zeta_{n+1}) \geq 1$ for all $n \in \mathbb{N} \cup \{0\}$. As in the proof of Theorem 2.1 of [1] we obtain $\rho(\zeta_n, \zeta_{n+1}) \leq \psi^n(\rho(\zeta_0, \zeta_1))$ and deduce that $(\zeta_n)_{n\in\mathbb{N}}$ is a left K-Cauchy sequence in $(\mathcal{X}, \rho)$ (see [1] (p. 2156)). Since $(\mathcal{X}, \rho)$ is left K-complete there exists $\theta \in \mathcal{X}$ such that $\rho(\theta, \zeta_n) \to 0$ as $n \to \infty$. From property (A) it follows that $\alpha(\theta, \zeta_n) \geq 1$ for all $n \in \mathbb{N} \cup \{0\}$. We shall show that θ is a fixed point of $\mathcal{T}$. Indeed, for each $n \in \mathbb{N} \cup \{0\}$ we have: $\rho(\mathcal{T}\theta, \zeta_{n+1}) = \rho(\mathcal{T}\theta, \mathcal{T}\zeta_n) \leq \alpha(\theta, \zeta_n)\rho(\mathcal{T}\theta, \mathcal{T}\zeta_n) \leq \psi(\rho(\theta, \zeta_n))$.

Since $\rho(\theta, \zeta_n) > 0$, we deduce that $\psi(\rho(\theta, \zeta_n)) < \rho(\theta, \zeta_n)$ (see e.g., [1] (Lemma 2.1)), and, hence, $\rho(\mathcal{T}\theta, \zeta_n) \to 0$ as $n \to \infty$. Since $(\mathcal{X}, \rho)$ is Hausdorff we conclude that $\theta = \mathcal{T}\theta$. □

As for metric spaces [1] (Theorem 2.1), a slight modification of the proof of Theorem 1 shows the following result where the property (A) is replaced by continuity of $\mathcal{T}$. More precisely we have

Theorem 2. *Let $(\mathcal{X}, \rho)$ be a Hausdorff left K-complete quasi-metric space and $\mathcal{T} : \mathcal{X} \to \mathcal{X}$ be an α–ψ-contractive mapping such that*

(i) *$\mathcal{T}$ is α-admissible;*
(ii) *there exists $\zeta_0 \in \mathcal{X}$ such that $\alpha(\zeta_0, \mathcal{T}\zeta_0) \geq 1$;*
(iii) *$\mathcal{T}$ is continuous.*

Then $\mathcal{T}$ has a fixed point.

Theorems 1 and 2 can not be generalized to $\mathcal{T}_1$ left K-complete quasi-metric spaces (see e.g., [23] (Example 5)).

Let us recall that if ρ is a quasi-metric on a set $\mathcal{X}$, then the function ρ^s defined on $\mathcal{X} \times \mathcal{X}$ by $\rho^s(\zeta, \eta) = \max\{\rho(\zeta, \eta), \rho(\eta, \zeta)\}$ is a metric on $\mathcal{X}$. We give an example for a quasi-metric space $(\mathcal{X}, \rho)$ where we can apply both Theorem 1 and Theorem 2 but not [1] (Theorem 2.2) because the metric space $(\mathcal{X}, \rho^s)$ is not complete.

Example 1. *Let* $\mathcal{X} := \{0\} \cup \{1/n : n \in \mathbb{N}\} \cup \{n : n \in \mathbb{N} \setminus \{1\}\}$. *It is routine to check that* $(\mathcal{X}, \rho)$ *is a Hausdorff quasi-metric space where (the quasi-metric) ρ is defined as follows:*

$\rho(\zeta, \zeta) = 0$ *for all* $\zeta \in \mathcal{X}$.
$\rho(0, 1/n) = 1/n$ *for all* $n \in \mathbb{N}$.
$\rho(1/n, 1/m) = 1/n$ *whenever* $n < m$.
$\rho(0, n) = 2^{-n}$ *for all* $n \in \mathbb{N} \setminus \{1\}$.
$\rho(n, m) = |2^{-n} - 2^{-m}|$ *for all* $n, m \in \mathbb{N} \setminus \{1\}$, *and*
$\rho(\zeta, \eta) = 1$ *otherwise.*

Observe that $(\mathcal{X}, \rho)$ *is left K-complete: The sequence* $(1/n)_{n \in \mathbb{N}}$ *is left K-Cauchy and converges to 0, whereas the sequence* $(n)_{n \in \mathbb{N}}$ *is Cauchy in the metric space* $(\mathcal{X}, \rho^s)$, *and hence left K-Cauchy in* $(\mathcal{X}, \rho)$, *and also converges to 0. However, we have* $\rho(n, 0) = 1$ *for all* $n \in \mathbb{N}$, *and thus the metric space* $(\mathcal{X}, \rho^s)$ *is not complete.*

Now define $\mathcal{T} : \mathcal{X} \to \mathcal{X}$ *as* $\mathcal{T}0 = 0$, $\mathcal{T}n = n+1$ *for all* $n \in \mathbb{N}$, *and* $\mathcal{T}(1/n) = n$ *for all* $n \in \mathbb{N} \setminus \{1\}$.

We show that $\mathcal{T}$ *is an* α–ψ-SVV *contractive mapping for* α *given by* $\alpha(0, n) = \alpha(n, n+1) = 1$ *for all* $n \in \mathbb{N}$, *and* $\alpha(\zeta, \eta) = 0$ *otherwise; and* $\psi \in \Psi$ *given by* $\psi(t) = t/2$ *for all* $t \geq 0$.

Indeed, since $\alpha(1, T1) = \alpha(1, 2) = 1$, *we deduce by the definition of* $\mathcal{T}$ *and the construction of* α *that* $\mathcal{T}$ *is* α-*contractive. Also, the property (A) is clearly satisfied since* $\rho(0, n) \to 0$ *as* $n \to \infty$, *and* $\alpha(0, n) = 1$ *for all* $n \in \mathbb{N}$. *It remains to check that* $\mathcal{T}$ *is an* α–ψ-*contractive mapping. To this end, it suffices to consider the following two cases:*

Case 1. $\zeta = 0, \eta = n, n \in \mathbb{N}$. *Thus, we obtain*

$$\alpha(\zeta, \eta)\rho(T\zeta, T\eta) = \alpha(0, n)\rho(0, n+1) = 2^{-(n+1)} \leq \frac{1}{2}\rho(0, n) = \psi(\rho(\zeta, \eta)).$$

Case 2. $\zeta = n, \eta = n+1, n \in \mathbb{N}$. *Thus, we obtain*

$$\begin{aligned}\alpha(\zeta, \eta)\rho(T\zeta, T\eta) &= \alpha(n, n+1)\rho(Tn, T(n+1)) = \rho(n+1, n+2) \\ &= 2^{-(n+2)} = \frac{1}{2}\rho(n, n+1) = \psi(\rho(\zeta, \eta)).\end{aligned}$$

Therefore, all conditions of Theorem 1 are satisfied.
Clearly, we can also apply Theorem 2 because $\mathcal{T}$ *is continuous (with respect to* τ_ρ).

Now, we present an easy example where we can apply Theorem 1 but not Theorem 2.

Example 2. *Let* $\mathcal{X} := \{0, \infty\} \cup \mathbb{N}$. *Clearly,* $(\mathcal{X}, \rho)$ *is a Hausdorff left K-complete quasi-metric space where (the quasi-metric) ρ is defined as follows:*

$\rho(\zeta, \zeta) = 0$ *for all* $\zeta \in \mathcal{X}$.
$\rho(0, 1/n) = 1/n$ *for all* $n \in \mathbb{N}$, *and*
$\rho(\zeta, \eta) = 1$ *otherwise.*

Now define $\mathcal{T} : \mathcal{X} \to \mathcal{X}$ *as* $\mathcal{T}0 = 0$, $\mathcal{T}\infty = \infty$, *and* $\mathcal{T}n = \infty$ *for all* $n \in \mathbb{N}$.

Since $\rho(0,n) \to 0$ as $n \to \infty$, but $\rho(\mathcal{T}0, \mathcal{T}n) = \rho(0, \infty) = 1$, we conclude that $\mathcal{T}$ is not continuous. However, it is obvious that $\mathcal{T}$ is an α–ψ-SVV contractive mapping for α given by $\alpha(\infty, \infty) = 1$, and $\alpha(\zeta, y) = 0$ otherwise, and any $\psi \in \Psi$.

In our main result (Theorem 3 below), we prove that Theorem 1 characterizes left K-completeness of Hausdorff quasi-metric spaces. However, Theorem 2 does not provide such characterization even in the case of metric spaces, as Suzuki and Takahashi constructed in [24] an example of a non-complete metric space for which every continuous self map has fixed points.

Theorem 3. *A Hausdorff quasi-metric space is left K-complete if and only if every α–ψ-SVV contractive mapping has a fixed point.*

Proof of Theorem 3. Let $(\mathcal{X}, \rho)$ be a Hausdorff left K-complete quasi-metric space. By Theorem 1, every α–ψ-SVV contractive mapping on $(\mathcal{X}, \rho)$ has a fixed point.

Conversely, suppose that $(\mathcal{X}, \rho)$ is a Hausdorff quasi-metric space which is not left K-complete. Then there exists a left K-Cauchy sequence $(\zeta_n)_{n\in\mathbb{N}}$ (of distinct points) in (ζ, ρ) which is not convergent for τ_ρ. Put $\mathcal{A} = \{\zeta_n : n \in \mathbb{N}\}$. Since $\rho(\zeta_1, \mathcal{A}\setminus\{\zeta_1\}) > 0$, there exists $h_1 \in \mathbb{N}$, with $h_1 > 1$, such that $\rho(\zeta_j, \zeta_k) < \rho(\zeta_1, \mathcal{A}\setminus\{\zeta_1\})/2$ whenever $h_1 \leq j \leq k$. Similarly, there exists $h_2 \in \mathbb{N}$, with $h_2 > \max\{2, h_1\}$, such that $\rho(\zeta_j, \zeta_k) < \rho(\zeta_2, \mathcal{A}\setminus\{\zeta_2\})/2$ whenever $h_2 \leq j \leq k$. In this way we obtain a subsequence $(h_n)_{n\in\mathbb{N}}$ of $(n)_{n\in\mathbb{N}}$ such that $h_n > \max\{n, h_{n-1}\}$ and $\rho(\zeta_j, \zeta_k) < \rho(\zeta_n, \mathcal{A}\setminus\{\zeta_n\})/2$ whenever $h_n \leq j \leq k$.

Define $\mathcal{T} : \mathcal{X} \to \mathcal{X}$ and $\alpha : \mathcal{X} \times \mathcal{X} \to [0, \infty)$ as follows:

$\mathcal{T}\zeta_n = \zeta_{h_n}$ for $n \in \mathbb{N}$, and $\mathcal{T}\zeta = \zeta_1$ for $\zeta \in \mathcal{X}\setminus\mathcal{A}$, and

$\alpha(\zeta, \eta) = 1$ if $\zeta = \zeta_n$ and $\eta = \zeta_m$ for $n, m \in \mathbb{N}$ with $n < m$, and $\alpha(\zeta, \eta) = 0$ otherwise.

We first note that $\alpha(\zeta_1, \mathcal{T}\zeta_1) = 1$ because $1 < h_1$.

Moreover $\mathcal{T}$ is α-admissible. Indeed, if $\alpha(\zeta, \eta) \geq 1$, then $\zeta = \zeta_n$ and $\eta = \zeta_m$ with $n < m$. So $\alpha(\mathcal{T}\zeta, \mathcal{T}\eta) = \alpha(\zeta_{h_n}, \zeta_{h_m}) = 1$ because $h_n < h_m$.

Next, we show that $\mathcal{T}$ is α-ψ-contractive for $\psi \in \Psi$ given by $\psi(t) = t/2$. Indeed, by the construction of α it suffices to check the case that $\zeta = \zeta_n$ and $\eta = \zeta_m$ with $n < m$. Thus, we obtain

$$\begin{aligned}\alpha(\zeta, \eta)\rho(\mathcal{T}\zeta, \mathcal{T}\eta) &= \alpha(\zeta_n, \zeta_m)\rho(\zeta_{h_n}, \zeta_{h_m}) < \frac{1}{2}\rho(\zeta_n, \mathcal{A}\setminus\{\zeta_n\}) \\ &\leq \frac{1}{2}\rho(\zeta_n, \zeta_m) = \frac{1}{2}\rho(\zeta, \eta) = \psi(\rho(\zeta, \eta)).\end{aligned}$$

Finally, note that $(\mathcal{X}, \rho)$ trivially satisfies the property (A) because the only convergent sequences in $\mathcal{A}$ are those that are eventually constant.

We have shown that $\mathcal{T}$ is an α–ψ-SVV contractive mapping without fixed point. This contradiction concludes the proof. □

Corollary 1. *A metric space is complete if and only every α–ψ-SVV contractive mapping has a fixed point.*

Author Contributions: Investigation, S.R. and P.T.; Writing–original draft, S.R. and P.T. All authors contributed equally in writing this article. All authors have read and agreed to the published version of the manuscript.

Funding: This research was partially funded by Ministerio de Ciencia, Innovación y Universidades, under grant PGC2018-095709-B-C21 and AEI/FEDER, UE funds.

Acknowledgments: The authors thank the reviewers for their useful suggestions and comments.

Conflicts of Interest: The authors declare no conflict of interest.

References

1. Samet, B.; Vetro, C.; Vetro, P. Fixed point theorems for α–ψ-contractive type mappings. *Nonlinear Anal.* **2012**, *75*, 2154–2165. [CrossRef]
2. Bhaskar, T.G.; Lakshmikantham, V. Fixed point theorems in partially ordered metric spaces and applications. *Nonlinear Anal.* **2006**, *65*, 1379–1393. [CrossRef]
3. Nieto, J.J.; Rodríguez-López, R. Contractive mapping theorems in partially ordered sets and applications to ordinary differential equations. *Order* **2005**, *22*, 223–239. [CrossRef]
4. Nieto, J.J.; Rodríguez-López, R. Existence and uniqueness of fixed point in partially ordered sets and applications to ordinary differential equations. *Act. Math. Sin. (Engl. Ser.)* **2007**, *23*, 2205–2212. [CrossRef]
5. Ran, A.C.M.; Reurings, M.C.B. A fixed point theorem in partially ordered sets and some applications to matrix equations. *Proc. Am. Math. Soc.* **2003**, *132*, 1435–1443. [CrossRef]
6. Amiri, P.; Rezapur, S.; Shahzad, N. Fixed points of generalized α–ψ-contractions. *Rev. R. Acad. Cienc. Exactas Fis. Nat. Ser. A. Mat. RACSAM* **2014**, *108*, 519–526. [CrossRef]
7. Bilgili, N.; Karapinar, E.; Samet, B. Generalized α–ψ contractive mappings in quasi-metric spaces and related fixed-point theorems. *J. Inequal. Appl.* **2014**, *2014*, 36. [CrossRef]
8. Bota, M.; Chifu, C.; Karapinar, E. Fixed point theorems for generalized $(\alpha$–$\psi)$-Ciric-type contractive multivalued operators in b-metric spaces. *J. Nonlinear Sci. Appl.* **2016**, *9*, 1165–1177. [CrossRef]
9. Karapinar, E. α–ψ-Geraghty contraction type mappings and some related fixed point results. *Filomat* **2014**, *28*, 37–48. [CrossRef]
10. Karapinar, E.; Dehici, A.; Redje, N. On some fixed points of α–ψ contractive mappings with rational expressions. *J. Nonlinear Sci. Appl.* **2017**, *10*, 1569–1581. [CrossRef]
11. Shahi, P.; Kaur, J.; Bhatia, S.S. Coincidence and common fixed point results for generalized α–ψ contractive type mappings with applications. *Bull. Belg. Math. Soc.* **2015**, *22*, 299–318. [CrossRef]
12. Mlaiki, N.; Kukić, K.; Gardašević-Filipović, M.; Aydi, H. On almost b-metric spaces and related fixed points results. *Axioms* **2019**, *8*, 70. [CrossRef]
13. Fulga, A.; Taş, A. Fixed point results via simulation functions in the context of quasi-metric space. *Filomat* **2018**, *32*, 4711–4729. [CrossRef]
14. Hu, T.K. On a fixed point theorem for metric spaces. *Amer. Math. Mon.* **1967**, *74*, 436–437. [CrossRef]
15. Kirk, A.W. Caristi's fixed point theorem and metric convexity. *Colloq. Math.* **1976**, *36*, 81–86. [CrossRef]
16. Subrahmanyan, P.V. Completeness and fixed-points. *Mon. Math.* **1975**, *80*, 325–330. [CrossRef]
17. Suzuki, T. A generalized Banach contraction principle that characterizes metric completeness. *Proc. Am. Math. Soc.* **2008**, *136*, 1861–1869. [CrossRef]
18. Romaguera, S. A Kirk type characterization of completeness for partial metric spaces. *Fixed Point Theory Appl.* **2010**, *2010*, 493298. [CrossRef]
19. Altun, I.; Romaguera, S. Characterizations of partial metric completeness in terms of weakly contractive mappings having fixed point. *Appl. Anal. Discr. Math.* **2012**, *6*, 247–256. [CrossRef]
20. Romaguera, S.; Tirado, P. A characterization of Smyth complete quasi-metric spaces via Caristi's fixed point theorem. *Fixed Point Theory Appl.* **2015**, *2015*, 183. [CrossRef]
21. Alegre, C.; Dağ, H.; Romaguera, S.; Tirado, P. Characterizations of quasi-metric completeness in terms of Kannan-type fixed point theorems. *Hacet. J. Math. Stat.* **2017**, *46*, 67–76. [CrossRef]
22. Cobzaş, S. *Functional Analysis in Asymmetric Normed Spaces*; Frontiers in Mathematics; Birkhäuser/Springer Basel AG: Basel, Switzerland, 2013.
23. Romaguera, S.; Tirado, P. The Meir-Keeler fixed point theorem for quasi-metric spaces and some consequences. *Symmetry* **2019**, *11*, 741. [CrossRef]
24. Suzuki, T.; Takahashi, W. Fixed point theorems and characterizations of metric completeness. *Topolog. Meth. Nonlinear Anal.* **1996**, *8*, 371–382. [CrossRef]

© 2019 by the authors. Licensee MDPI, Basel, Switzerland. This article is an open access article distributed under the terms and conditions of the Creative Commons Attribution (CC BY) license (http://creativecommons.org/licenses/by/4.0/).

Article

A Novel Delay-Dependent Asymptotic Stability Conditions for Differential and Riemann-Liouville Fractional Differential Neutral Systems with Constant Delays and Nonlinear Perturbation

Watcharin Chartbupapan [1], Ovidiu Bagdasar [2] and Kanit Mukdasai [1],*

1. Department of Mathematics, Faculty of Science, Khon Kaen University, khon Kaen 40002, Thailand; chartbupapan@gmail.com
2. Department of Electronics, Computing and Mathematics, University of Derby, Derby DE22 1GB, UK; o.bagdasar@derby.ac.uk
* Correspondence: kanit@kku.ac.th

Received: 28 November 2019; Accepted: 23 December 2019; Published: 3 January 2020

Abstract: The novel delay-dependent asymptotic stability of a differential and Riemann-Liouville fractional differential neutral system with constant delays and nonlinear perturbation is studied. We describe the new asymptotic stability criterion in the form of linear matrix inequalities (LMIs), using the application of zero equations, model transformation and other inequalities. Then we show the new delay-dependent asymptotic stability criterion of a differential and Riemann-Liouville fractional differential neutral system with constant delays. Furthermore, we not only present the improved delay-dependent asymptotic stability criterion of a differential and Riemann-Liouville fractional differential neutral system with single constant delay but also the new delay-dependent asymptotic stability criterion of a differential and Riemann-Liouville fractional differential neutral equation with constant delays. Numerical examples are exploited to represent the improvement and capability of results over another research as compared with the least upper bounds of delay and nonlinear perturbation.

Keywords: asymptotic stability; differential and riemann-liouville fractional differential neutral systems; linear matrix inequality

1. Introduction

Differential systems, or more generally functional differential systems, have been studied rather extensively for at least 200 years and are used as models to describe transportation systems, communication networks, teleportation systems, physical systems and biological systems, and so forth. Parts of fractional-order systems have not received much attention by reason of absence of appropriate utilization circumstances over the past 300 years. However, during the last 10 years fractional-order systems have been widely investigated as they have the qualification to explain various phenomena more precisely in many fields, for example, biological models, material science, finance, cardiac tissues, quantum mechanics, viscoelastic systems, medicine and fluid mechanics [1–8]. Caputo fractional differential systems have been studied in many types of stability such as uniform stability [9], Mittag-Leffler stability [10–13], Ulam stability [14], finite time stability [15,16] and asymptotic stability [17,18]. Nevertheless, the stability of Riemann-Liouville fractional differential systems is seldom considered, see References [19,20].

The neutral systems with time delays have already been applied in many fields, such as heartbeat, memorization, locomotion, mastication and respiration, see References [21–24]. Accordingly, the issue of stability analysis for differential and Riemann-Liouville fractional differential neutral systems has attracted researchers. The asymptotic stability criteria for certain neutral differential equations (CNDE) with constant delays have been discussed in References [25–29] by applying Lyapunov-Krasovskii functional and several model transformations. In References [30–33], the researchers considered the exponential stability problem for CNDE with time-varying delays by several methods. In Reference [30], the results were established without the use of the bounding technique and the model transformation method, while researchers have studied it by using radially unboundedness, the Lyapunov-Krasovskii functional approach and the model transformation method in Reference [32]. Moreover, in Reference [34] Li et al. presented the asymptotic stability conditions for fractional neutral systems in the form of matrix measure and matrix norm of the system matrices. However, the criteria, drafted in the form of matrix norm, are more conservative, while Liu et al. used the Lyapunov direct method to establish the asymptotic stability criteria of Riemann-Liouville fractional neutral systems in the form of LMIs [35].

This paper is involved with the analysis problem for the asymptotic stability of differential and Riemann-Liouville fractional differential neutral systems with constant delays and nonlinear perturbation by applying a zero equation, model transformation and other inequalities. The novel asymptotic stability condition is instituted in the form of LMIs. Then we show the new delay-dependent asymptotic stability criterion of differential and Riemann-Liouville fractional differential neutral systems with constant delays. In addition, the improved delay-dependent asymptotic stability criterion of differential and Riemann-Liouville fractional differential neutral systems with single constant delay and the new delay-dependent asymptotic stability criterion of differential and Riemann-Liouville fractional differential neutral equations with constant delays are established. Numerical examples represent the capability of our results as compared with other research.

2. Problem Formulation and Preliminaries

We introduce a differential and fractional differential neutral system with constant delays and nonlinear perturbation

$$_{t_0}D_t^q[x(t) + Cx(t-\tau)] = -Ax(t) + Bx(t-\sigma) + f(x(t-\sigma)), \quad t > 0, \tag{1}$$
$$x(t) = \varrho(t), \quad t \in [-\kappa, 0],$$

for $0 < q \leq 1$, the state vector $x(t) \in \mathbb{R}^n$, A, B, C are symmetric positive definite matrices with $\|C\| < 1$, τ, σ are positive real constants and $\varrho \in C([-\kappa, 0]; \mathbb{R}^n)$ with $\kappa = \max\{\tau, \sigma\}$.

The uncertainty $f(.)$ represents the nonlinear parameter perturbation satisfying

$$f^T(x(t))f(x(t)) \leq \delta^2 x^T(t)x(t), \tag{2}$$
$$f^T(x(t-\sigma))f(x(t-\sigma)) \leq \eta^2 x^T(t-\sigma)x(t-\sigma), \tag{3}$$

where δ, η are given constants.

Next, the Riemann-Liouville fractional integral and derivative [36] are defined as, respectively

$$_{t_0}D_t^{-q}x(t) = \frac{1}{\Gamma(q)} \int_{t_0}^t (t-s)^{q-1}x(s)ds, \quad (q > 0), \tag{4}$$

$$_{t_0}D_t^q x(t) = \frac{1}{\Gamma(n-q)} \frac{d^n}{dt^n} \int_{t_0}^t \frac{x(s)}{(t-s)^{q+1-n}}ds, \quad (n-1 \leq q < n). \tag{5}$$

Lemma 1. *[37] For $x(t) \in \mathbb{R}^n$ and $p > q > 0$, then*

$$_{t_0}D_t^q(_{t_0}D_t^{-p}x(t)) = {}_{t_0}D_t^{q-p}x(t). \tag{6}$$

Lemma 2. *[17] For a vector of differentiable function $x(t) \in \mathbb{R}^n$, positive semi-definite matrix $K \in \mathbb{R}^{n \times n}$ and $0 < q < 1$, then*

$$\frac{1}{2}{}_{t_0}D_t^q(x^T(t)Kx(t)) \leq x^T(t)K{}_{t_0}D_t^q x(t), \tag{7}$$

for all $t \geq t_0$.

3. Main Results

Consider the asymptotic stability for system (1) with constant delays and nonlinear perturbation. We define a new variable

$$\Psi(t) = x(t) + Cx(t - \tau). \tag{8}$$

Rewrite the Equation (1) in the following equation

$${}_{t_0}D_t^q \Psi(t) = -Ax(t) + Bx(t - \sigma) + f(x(t - \sigma)). \tag{9}$$

Theorem 1. *Let δ and η be positive scalars, if there are any appropriate dimensions matrices $Q_j(j = 1, 2, 3)$ and symmetric positive definite matrices $K_i(i = 1, 2, 3, 4, 5)$ such that satisfy*

$$\Sigma = \begin{bmatrix} -Q_1 - Q_1^T & \Omega_{(1,2)} & Q_1 C - Q_3^T & K_1 & K_1 B \\ * & \Omega_{(2,2)} & Q_2 C + Q_3^T & 0 & 0 \\ * & * & \Omega_{(3,3)} & 0 & 0 \\ * & * & * & -K_5 - \sigma I & 0 \\ * & * & * & * & -K_3 + \sigma \eta^2 I \end{bmatrix} < 0, \tag{10}$$

where
$\Omega_{(1,2)} = -K_1 A + Q_1 - Q_2^T,$
$\Omega_{(2,2)} = Q_2 + Q_2^T + K_2 + K_3 + \tau K_4 + \delta^2 K_5,$
$\Omega_{(3,3)} = Q_3 C + C^T Q_3^T - K_2.$

Then the system (1) is asymptotically stable.

Proof of Theorem 1. For symmetric positive definite matrices $K_i(i = 1, 2, 3, 4, 5)$ and any appropriate dimensions matrices $Q_j(j = 1, 2, 3)$. Consider the Lyapunov-Krasovskii functional

$$V(t) = \sum_{i=1}^{2} V_i(t), \tag{11}$$

for

$$\begin{aligned} V_1(t) &= {}_{t_0}D_t^{q-1} \Psi^T(t) K_1 \Psi(t), \\ V_2(t) &= \int_{t-\tau}^{t} x^T(s) K_2 x(s) ds + \int_{t-\sigma}^{t} x^T(s) K_3 x(s) ds \\ &\quad + \int_{t-\tau}^{t} (\tau - t + s) x^T(s) K_4 x(s) ds \\ &\quad + \int_{t-\sigma}^{t} f^T(x(s)) K_5 f(x(s)) ds. \end{aligned}$$

Computing the differential of $V(t)$ on the solution of system (1)

$$\dot{V}(t) = \sum_{i=1}^{2} \dot{V}_i(t). \tag{12}$$

The differential of $V_1(t)$ is computed by Lemma 2

$$\begin{aligned}
\dot{V}_1(t) &= {}_{t_0}D_t^q \Psi^T(t) K_1 \Psi(t) \\
&\leq 2\Psi^T(t) K_1 ({}_{t_0}D_t^q \Psi(t)) \\
&= 2\Psi^T(t) K_1 [-Ax(t) + Bx(t-\sigma) + f(x(t-\sigma))] \\
&\quad + 2\Psi^T(t) Q_1 [-\Psi(t) + x(t) + Cx(t-\tau)] \\
&\quad + 2x^T(t) Q_2 [-\Psi(t) + x(t) + Cx(t-\tau)] \\
&\quad + 2x^T(t-\tau) Q_3 [-\Psi(t) + x(t) + Cx(t-\tau)].
\end{aligned} \tag{13}$$

Taking the differential of $V_2(t)$, we obtain

$$\begin{aligned}
\dot{V}_2(t) &= x^T(t) K_2 x(t) - x^T(t-\tau) K_2 x(t-\tau) \\
&\quad + x^T(t) K_3 x(t) - x^T(t-\sigma) K_3 x(t-\sigma) \\
&\quad + \tau x^T(t) K_4 x(t) - \int_{t-\tau}^{t} x^T(s) K_4 x(s) \\
&\quad + f^T(x(t)) K_5 f(x(t)) - f^T(x(t-\sigma)) K_5 f(x(t-\sigma)) \\
&\leq x^T(t) K_2 x(t) - x^T(t-\tau) K_2 x(t-\tau) \\
&\quad + x^T(t) K_3 x(t) - x^T(t-\sigma) K_3 x(t-\sigma) \\
&\quad + \tau x^T(t) K_4 x(t) + \delta^2 x^T(t) K_5 x(t) \\
&\quad - f^T(x(t-\sigma)) K_5 f(x(t-\sigma)).
\end{aligned} \tag{14}$$

Next, from (3), we obtain

$$0 \leq \sigma \eta^2 x^T(t-\sigma) x(t-\sigma) - \sigma f^T(x(t-\sigma)) f(x(t-\sigma)). \tag{15}$$

According to (13), (14) and (15), we can conclude that

$$\dot{V}(t) \leq \xi^T(t) \sum \xi(t), \tag{16}$$

where $\xi(t) = \text{col}\{\Psi(t), x(t), x(t-\tau), f(x(t-\sigma)), x^T(t-\sigma)\}$.

Since linear matrix inequality (10) holds, then the system (1) is asymptotic stability. □

Next, we consider system (1) with $f(x(t-\sigma)) = 0$,

$$\begin{aligned}
{}_{t_0}D_t^q [x(t) + Cx(t-\tau)] &= -Ax(t) + Bx(t-\sigma) \quad t > 0, \\
x(t) &= \varrho(t), \quad t \in [-\kappa, 0],
\end{aligned} \tag{17}$$

for $0 < q \leq 1$, the state vector $x(t) \in \mathbb{R}^n$, A, B, C are symmetric positive definite matrices with $\|C\| < 1$, τ, σ are positive real constants and $\varrho \in C([-\kappa, 0]; \mathbb{R}^n)$ with $\kappa = \max\{\tau, \sigma\}$.

Corollary 1. *If there are any appropriate dimensions matrices $Q_j (j = 1, 2, 3)$ and symmetric positive definite matrices $K_i (i = 1, 2, 3, 4)$ such that satisfy*

$$\begin{bmatrix} -Q_1 - Q_1^T & -K_1 A + Q_1 - Q_2^T & Q_1 C - Q_3^T & K_1 B \\ * & Q_2 + Q_2^T + K_2 + K_3 + \tau K_4 & Q_2 C + Q_3^T & 0 \\ * & * & Q_3 C + C^T Q_3^T - K_2 & 0 \\ * & * & * & -K_3 \end{bmatrix} < 0. \tag{18}$$

Then the system (17) is asymptotically stable.

Proof of Corollary 1. For symmetric positive definite matrices $K_i(i = 1,2,3,4)$ and any appropriate dimensions matrices $Q_j(j = 1,2,3)$. Consider the Lyapunov-Krasovskii functional

$$V(t) = \sum_{i=1}^{2} V_i(t), \tag{19}$$

for

$$\begin{aligned}
V_1(t) &= {}_{t_0}D_t^{q-1}\Psi^T(t)K_1\Psi(t), \\
V_2(t) &= \int_{t-\tau}^{t} x^T(s)K_2 x(s)ds + \int_{t-\sigma}^{t} x^T(s)K_3 x(s)ds \\
&\quad + \int_{t-\tau}^{t} (\tau - t + s)x^T(s)K_4 x(s)ds.
\end{aligned}$$

According to Theorem 1, we present the asymptotic stability criterion (18) of system (17). □

Next, we consider system (1) with $f(x(t - \sigma)) = 0$ and $\sigma = \tau$,

$$\begin{aligned}
{}_{t_0}D_t^q[x(t) + Cx(t - \tau)] &= -Ax(t) + Bx(t - \tau) \quad t > 0, \\
x(t) &= \varrho(t), \quad t \in [-\tau, 0],
\end{aligned} \tag{20}$$

for $0 < q \leq 1$, the state vector $x(t) \in \mathbb{R}^n$, A, B, C are symmetric positive definite matrices with $\|C\| < 1$, τ is positive real constants and $\varrho \in C([-\tau, 0]; \mathbb{R}^n)$.

Corollary 2. *If there are any appropriate dimensions matrices $Q_j(j = 1,2,3)$ and symmetric positive definite matrices $K_i(i = 1,2,3)$ such that satisfy*

$$\begin{bmatrix} -Q_1 - Q_1^T & -K_1 A + Q_1 - Q_2^T & Q_1 C - Q_3^T + K_1 B \\ * & Q_2 + Q_2^T + K_2 + \tau K_3 & Q_2 C + Q_3^T \\ * & * & Q_3 C + C^T Q_3^T - K_2 \end{bmatrix} < 0. \tag{21}$$

Then the Equation (20) is asymptotically stable.

Proof of Corollary 2. For symmetric positive definite matrices $K_i(i = 1,2,3)$ and any appropriate dimensions matrices $Q_j(j = 1,2,3)$. Consider the Lyapunov-Krasovskii functional

$$V(t) = \sum_{i=1}^{2} V_i(t), \tag{22}$$

for

$$\begin{aligned}
V_1(t) &= {}_{t_0}D_t^{q-1}\Psi^T(t)K_1\Psi(t), \\
V_2(t) &= \int_{t-\tau}^{t} x^T(s)K_2 x(s)ds \\
&\quad + \int_{t-\tau}^{t} (\tau - t + s)x^T(s)K_3 x(s)ds.
\end{aligned} \tag{23}$$

According to Theorem 1, we present the asymptotic stability criterion (21) of system (20). □

4. Application

$$_{t_0}D_t^q[x(t) + px(t-\tau)] = -ax(t) + b\tanh x(t-\sigma) \quad t > 0, \quad (24)$$
$$x(t) = \varrho(t), \quad t \in [-\kappa, 0],$$

for $0 < q \leq 1$, the state vector $x(t) \in \mathbb{R}$, a, b, p are real constants with $|p| < 1$, τ, σ are positive real constants $\varrho \in C([-\kappa, 0]; \mathbb{R})$ with $\kappa = \max\{\tau, \sigma\}$.

Corollary 3. *If there are positive real constants $k_i (i = 1, 2, 3, 4, 5)$ and real constants $q_j (j = 1, 2, 3)$ such that satisfy*

$$\begin{bmatrix} -2q_1 & -k_1a + q_1 - q_2 & q_1p - q_3 & k_1b & 0 \\ * & 2q_2 + k_2 + k_3 + k_4\tau + k_5 & q_2p + q_3 & 0 & 0 \\ * & * & 2q_3p - k_2 & 0 & 0 \\ * & * & * & -k_5 - \sigma & 0 \\ * & * & * & * & -k_3 + \sigma \end{bmatrix} < 0. \quad (25)$$

Then the Equation (24) is asymptotically stable.

Proof of Corollary 3. For positive real constants $k_i (i = 1, 2, 3, 4, 5)$ and real constants $q_j (j = 1, 2, 3)$. Consider the Lyapunov-Krasovskii functional

$$V(t) = \sum_{i=1}^{2} V_i(t), \quad (26)$$

for

$$V_1(t) = k_1{}_{t_0}D_t^{q-1}\Psi^2(t),$$
$$V_2(t) = k_2 \int_{t-\tau}^{t} x^2(s)ds + k_3 \int_{t-\sigma}^{t} x^2(s)ds$$
$$+ k_4 \int_{t-\tau}^{t} (\tau - t + s)x^2(s)ds + k_5 \int_{t-\sigma}^{t} \tanh x^2(s)ds.$$

According to Theorem 1, we present the asymptotic stability criterion (25) of system (3). □

5. Numerical Examples

Example 1. *The fractional neutral system :*

$$_{t_0}D_t^q[x(t) + Cx(t-0.5)] = -Ax(t) + Bx(t-\sigma) + f(x(t-\sigma)). \quad (27)$$

Solving the LMI (10) when $A = \begin{bmatrix} 1.45 & 0 \\ 0 & 1.45 \end{bmatrix}, B = \begin{bmatrix} 0 & 0.4 \\ 0.4 & 0 \end{bmatrix}, C = \begin{bmatrix} -0.1 & 0 \\ 0 & -0.1 \end{bmatrix}$, *we have a set of parameters that ensures asymptotic stability of system (27) which* $\eta = 5 \times 10^3$, $\delta = 1$ *and* $\sigma = 0.5$ *as follows:*

$K_1 = 10^8 \times \begin{bmatrix} 3.5993 & 0 \\ 0 & 3.5993 \end{bmatrix}$, $K_2 = 10^7 \times \begin{bmatrix} 1.3106 & 0 \\ 0 & 1.3106 \end{bmatrix}$, $K_3 = 10^8 \times \begin{bmatrix} 1.5730 & 0 \\ 0 & 1.5730 \end{bmatrix}$,

$K_4 = 10^6 \times \begin{bmatrix} 9.7620 & 0 \\ 0 & 9.7620 \end{bmatrix}$, $K_5 = 10^8 \times \begin{bmatrix} 3.5456 & 0 \\ 0 & 3.5456 \end{bmatrix}$, $Q_1 = 10^8 \times \begin{bmatrix} 2.9931 & 0 \\ 0 & 2.9931 \end{bmatrix}$,

$Q_2 = 10^8 \times \begin{bmatrix} -3.0980 & 0 \\ 0 & -3.0980 \end{bmatrix}$, $Q_3 = 10^7 \times \begin{bmatrix} -2.2267 & 0 \\ 0 & -2.2267 \end{bmatrix}$.

Moreover, the least upper bound of the parameter σ that ensures the asymptotic stability of system (27) is 1.3227 when $\eta = 5 \times 10^3$ and $\delta = 1$. Table 1 represents the least upper bound σ of this example for various values of η, δ.

Table 1. The least upper bound of σ for Example 1.

	$\eta = 5 \times 10^3$	$\eta = 6 \times 10^3$	$\eta = 7 \times 10^3$
$\delta = 0.8$	6.4920	4.5076	3.3117
$\delta = 0.9$	4.1166	2.8588	2.1003
$\delta = 1$	1.3227	0.9185	0.6748

Example 2. *The fractional neutral system :*

$$_{t_0}D_t^q[x(t) + Cx(t-\tau)] = -Ax(t) + Bx(t-1.2). \quad (28)$$

Solving the LMI (18) when $A = \begin{bmatrix} 1.45 & 0 \\ 0 & 1.45 \end{bmatrix}$, $B = \begin{bmatrix} 0 & 0.4 \\ 0.4 & 0 \end{bmatrix}$, $C = \begin{bmatrix} -0.1 & 0 \\ 0 & -0.1 \end{bmatrix}$, we have a set of parameters that ensures asymptotic stability of system (28) which $\tau = 0.6$ as follows:

$K_1 = \begin{bmatrix} 44.0782 & 0 \\ 0 & 44.0782 \end{bmatrix}$, $K_2 = \begin{bmatrix} 32.9861 & 0 \\ 0 & 32.9861 \end{bmatrix}$, $K_3 = \begin{bmatrix} 32.6501 & 0 \\ 0 & 32.6501 \end{bmatrix}$,

$K_4 = \begin{bmatrix} 31.8793 & 0 \\ 0 & 31.8793 \end{bmatrix}$, $Q_1 = \begin{bmatrix} 14.6090 & 0 \\ 0 & 14.6090 \end{bmatrix}$, $Q_2 = \begin{bmatrix} -56.7801 & 0 \\ 0 & -56.7801 \end{bmatrix}$,

$Q_3 = \begin{bmatrix} -3.3600 & 0 \\ 0 & -3.3600 \end{bmatrix}$.

Moreover, the least upper bound of the parameter τ that ensures the asymptotic stability of system (28) is 3.7×10^{22}.

Example 3. *The fractional neutral system :*

$$_{t_0}D_t^q[x(t) + Cx(t-\tau)] = -Ax(t) + Bx(t-\tau). \quad (29)$$

Solving the LMI (21) when $A = \begin{bmatrix} 3 & -1 \\ 0 & 1 \end{bmatrix}$, $B = \begin{bmatrix} 0.2 & 0.1 \\ 0 & 0.1 \end{bmatrix}$, $C = \begin{bmatrix} 0.1 & 0 \\ 0 & 0.2 \end{bmatrix}$, we obtain the least upper bound of the parameter τ that ensures the asymptotic stability is 2.86×10^{24}. By the criterion in [35], the least upper bound of the parameter τ is 2.99×10^{21}. This example represents our result is less conservative than these in [35].

Example 4. *The differential equation, which is considered in [25,27,30–32]:*

$$\frac{d}{dt}[x(t) + 0.35x(t-0.5)] = -1.5x(t) + b \tanh x(t-0.5). \quad (30)$$

By using linear matrix inequality (25), the comparison for the least upper bound b that ensures asymptotic stability of Equation (30) are represented in Table 2.

Table 2. The least upper bound of b for Example 4.

Deng et al. (2009) [25]	0.889
Nam and Phat (2009) [27]	1.405
Chen and Meng (2011) [31]	1.346
Chen (2012) [30]	1.405
Keadnarmol and Rojsiraphisal (2014) [32]	1.405
Corollary 3	1.4051

Example 5. *The differential equation in [27,30,31,38]:*

$$\frac{d}{dt}[x(t) + 0.2x(t-0.1)] = -0.6x(t) + 0.3\tanh x(t-\sigma). \tag{31}$$

By using linear matrix inequality (25), the comparison for the least upper bound delay σ that ensures asymptotic stability of Equation (31) are represented in Table 3.

Table 3. The least upper bound of σ for Example 5.

Nam and Phat (2009) [27]	2.32
Rojsiraphisal and Niamsup (2010) [38]	2.32
Chen and Meng (2011) [31]	10^{21}
Chen (2012) [30]	1.34×10^{21}
Corollary 3	6.21×10^{8}

Example 6. *The fractional neutral equation :*

$$_{t_0}D_t^q[x(t) + px(t-0.5)] = -ax(t) + b\tanh x(t-0.5). \tag{32}$$

Solving the LMI (25), we have a set of parameters that ensures asymptotic stability of Equation (32) which $a = 0.75, b = 0.3$ and $p = 0.4$ as follows:
$k_1 = 3.1544$, $k_2 = 1.0324$, $k_3 = 1.0749$, $k_4 = 0.7170$, $k_5 = 0.7385$, $q_1 = 0.7587$, $q_2 = -1.9721$, $q_3 = 0.4433$.

Furthermore, the least upper bound of b that ensures the asymptotic stability of Equation (32) is 0.6873 with $a = 0.75, p = 0.4$. Table 4 represents the least upper bound b of this example for various values of a, p.

Table 4. The least upper bound of b for Example 6.

	$a = 0.25$	$a = 0.5$	$a = 0.75$	$a = 1$	$a = 1.25$
$p = 0.2$	0.2449	0.4898	0.7348	0.9797	1.2247
$p = 0.4$	0.2291	0.4582	0.6873	0.9165	1.1456
$p = 0.6$	0.2000	0.3999	0.5999	0.7999	0.9999
$p = 0.8$	0.1500	0.2999	0.4499	0.5999	0.7499

6. Conclusions

The aim of this paper is a novel asymptotic stability analysis of differential and Riemann-Liouville fractional differential neutral systems with constant delays and nonlinear perturbation by applying zero equations, model transformation and other inequalities. The new asymptotic stability condition is given in the form of LMIs. Then we show the new delay-dependent asymptotic stability criterion of a differential and Riemann-Liouville fractional differential neutral system with constant delays. Furthermore, we propose the improved delay-dependent asymptotic stability criterion of differential and Riemann-Liouville fractional differential neutral systems with single constant delay and the new delay-dependent asymptotic stability criterion of differential and Riemann-Liouville fractional differential neutral equations with constant delays. Numerical examples illustrate the advantages and applicability of our results.

Author Contributions: All authors claim to have contributed significantly and equally to this work. All authors have read and agreed to the published version of the manuscript.

Funding: This work is supported by Science Achievement Scholarship of Thailand (SAST), Research and Academic Affairs Promotion Fund, Faculty of Science, Khon Kaen University, Fiscal year 2020 and National Research Council of Thailand and Khon Kaen University, Thailand (6200069).

Acknowledgments: The authors thank the reviewers for their valuable comments and suggestions, which led to the improvement of the content of the paper.

Conflicts of Interest: The authors declare no conflict of interest.

References

1. Ahmad, B.; Alghanmi, M.; Alsaedi, A.; Agarwal, R.V. Nonlinear impulsive multi-order Caputo-Type generalized fractional differential equations with infinite delay. *Mathematics* **2019**, *7*, 1108. [CrossRef]
2. Khan, U.; Ellahi, R.; Khan, R.; Mohyud-Din, S.T. Extracting new solitary wave solutions of Benny–Luke equation and Phi-4 equation of fractional order by using (G′/G)-expansion method. *Opt. Quant. Electron.* **2017**, *49*, 362. [CrossRef]
3. Lundstrom, B.N.; Higgs, M.H.; Spain, W.J.; Fairhall, A.L. Fractional differentiation by neocortical pyramidal neurons. *Nat. Neurosci.* **2008**, *11*, 1335–1342. [CrossRef] [PubMed]
4. Magin, R.L.; Ovadia, M. Modeling the cardiac tissue electrode interface using fractional calculus. *J. Vib. Control* **2008**, *14*, 1431–1442. [CrossRef]
5. Picozzi, S.; West, B.J. Fractional Langevin model of memory in financial markets. *Phys. Rev. E* **2002**, *66*, 46–118. [CrossRef] [PubMed]
6. Rahmatullah; Ellahi, R.; Mohyud-Din, S.T.; Khan, U. Exact traveling wave solutions of fractional order Boussinesq-like equations by applying Exp-function method. *Results Phys.* **2018**, *8*, 114–120. [CrossRef]
7. Sohail, A.; Maqbool, K.; Ellahi, R. Stability analysis for fractional-order partial differential equations by means of space spectral time Adams-Bashforth Moulton method. *Numer. Methods Partial. Differ. Equ.* **2017**, *34*, 19–29. [CrossRef]
8. Tripathi, D.; Pandey, S.K.; Das, S. Peristaltic flow of viscoelastic fluid with fractional Maxwell model through a channel. *Appl. Math. Comput.* **2010**, *215*, 3645–3654. [CrossRef]
9. Duarte-Mermoud, M.A.; Aguila-Camacho, N.; Gallegos, J.A.; Castro-Linares, R. Using general quadratic Lyapunov functions to prove Lyapunov uniform stability for fractional order systems. *Commun. Nonlinear Sci. Numer. Simul.* **2015**, *22*, 650–659. [CrossRef]
10. Chen, L.P.; He, Y.G.; Chai, Y.; Wu, R.C. New results on stability stabilization of a class of nonlinear fractional-order systems. *Nonlinear Dynam.* **2014**, *75*, 633–641. [CrossRef]
11. Wu, G.C.; Abdeljawad, T.; Liu, J.; Baleanu, D.; Wu, K.T. Mittag–Leffler stability analysis of fractional discrete–time neural networks via fixed point technique. *Nonlinear Anal. Model. Control* **2019**, *24*, 919–936. [CrossRef]
12. Wu, G.C.; Baleanu, D.; Huang, L.L. Novel Mittag-Leffler stability of linear fractional delay difference equations with impulse. *Appl. Math. Lett.* **2018**, *82*, 71–78. [CrossRef]
13. Yang, X.; Li, C.; Huang, T. Mittag-Leffler stability analysis of nonlinear fractional-order systems with impulses. *Appl. Math. Comput.* **2017**, *293*, 416–422. [CrossRef]
14. Brzdek, J.; Eghbali, N. On approximate solutions of some delayed fractional differential equations. *Appl. Math. Lett.* **2016**, *54*, 31–35. [CrossRef]
15. Chen, L.P.; Liu, C.; Wu, R.C.; He, Y.G.; Chai, Y. Finite-time stability criteria for a class of fractional-order neural networks with delay. *Neural Comput. Appl.* **2016**, *27*, 549–556. [CrossRef]
16. Li, M.; Wang, J.R. Finite time stability of fractional delay differential equations. *Appl. Math. Lett.* **2017**, *64*, 170–176. [CrossRef]
17. Liu, S.; Li, X.; Zhou, X.F.; Jiang, W. Lyapunov stability analysis of fractional nonlinear systems. *Appl. Math. Lett.* **2016**, *51*, 13–19. [CrossRef]
18. Liu, S.; Zhou, X.F.; Li, X.; Jiang, W. Stability of fractional nonlinear singular systems its applications in synchronization of complex dynamical networks. *Nonlinear Dynam.* **2016**, *84*, 2377–2385. [CrossRef]
19. Ozarslan, M.A.; Ustaoglu, C. Some incomplete hypergeometric functions and incomplete Riemann-Liouville fractional integral operators. *Mathematics* **2019**, *7*, 483. [CrossRef]

20. Rashid, S.; Abdeljawad, T.; Jarad, F.; Noor, M.A. Some estimates for generalized Riemann-Liouville fractional integrals of exponentially convex functions and their applications. *Mathematics* **2019**, *7*, 807. [CrossRef]
21. Fridman, E. Stability of linear descriptor systems with delays: a Lyapunov-based approach. *Aust. J. Math. Anal. Appl.* **2002**, *273*, 24–44. [CrossRef]
22. Kwon, O.M.; Park, J.H.; Lee, S.M. Augmented Lyapunov functional approach to stability of uncertain neutral systems with time-varying delays. *Appl. Comput. Math.* **2009**, *207*, 202–212. [CrossRef]
23. Liao, X.; Liu, Y.; Guo, S.; Mai, H. Asymptotic stability of delayed neural networks: a descriptor system approach. *Commun. Nonlinear Sci. Numer. Simul.* **2009**, *14*, 3120–3133. [CrossRef]
24. Park, J.H. Delay-dependent criterion for guaranteed cost control of neutral delay systems. *J. Opt. Theory Appl.* **2005**, *124*, 491–502. [CrossRef]
25. Deng, S.; Liao, X.; Guo, S. Asymptotic stability analysis of certain neutral differential equations: A descriptor system approach. *Math. Comput. Simul.* **2009**, *71*, 4297–4308. [CrossRef]
26. Kwon, O.M.; Park, J.H. On improved delay-dependent stability criterion of certain neutral differential equations. *Appl. Math. Comput.* **2008**, *199*, 385–391. [CrossRef]
27. Nam, P.T.; Phat, V.N. An improved stability criterion for a class of neutral differential equations. *Appl. Math. Lett.* **2009**, *22*, 31–35. [CrossRef]
28. Agarwal, R.P.; Grace, S.R. Asymptotic stability of certain neutral differential equations. *Math. Comput. Model.* **2000**, *31*, 9–15. [CrossRef]
29. Park, J.H.; Kwon, O.M. Stability analysis of certain nonlinear differential equation. *Chaos Solitons Fractals* **2008**, *27*, 450–453. [CrossRef]
30. Chen, H. Some improved criteria on exponential stability of neutral differential equation. *Adv. Differ. Equ.* **2012**, *2012*, 170. [CrossRef]
31. Chen, H.; Meng, X. An improved exponential stability criterion for a class of neutral delayed differential equations. *Appl. Math. Lett.* **2011**, *24*, 1763–1767. [CrossRef]
32. Keadnarmol, P.; Rojsiraphisal, T. Globally exponential stability of a certain neutral differential equation with time-varying delays. *Adv. Differ. Equ.* **2014**, *2014*, 32. [CrossRef]
33. Li, X. Global exponential stability for a class of neural networks. *Appl. Math. Lett.* **2009**, *22*, 1235–1239. [CrossRef]
34. Li, H.; Zhou, S.; Li, H. Asymptotic stability analysis of fractional-order neutral systems with time delay. *Adv. Differ. Equ.* **2015**, *2015*, 325–335. [CrossRef]
35. Liu, S.; Wu, X.; Zhang, Y.J.; Yang, R. Asymptotical stability of Riemann–Liouville fractional neutral systems. *Appl. Math. Lett.* **2017**, *69*, 168–173. [CrossRef]
36. Podlubny, I. *Fractional Differential Equations*; Academic Press: New York, NY, USA, 1999.
37. Kilbas, A.A.; Srivastava, H.M.; Trujillo, J.J. *Theory and Application of Fractional Differential Equations*; Elsevier: New York, NY, USA, 2006.
38. Rojsiraphisal, T.; Niamsup, P. Exponential stability of certain neutral differential equations. *Appl. Math. Lett.* **2010**, *17*, 3875–3880. [CrossRef]

© 2020 by the authors. Licensee MDPI, Basel, Switzerland. This article is an open access article distributed under the terms and conditions of the Creative Commons Attribution (CC BY) license (http://creativecommons.org/licenses/by/4.0/).

MDPI
St. Alban-Anlage 66
4052 Basel
Switzerland
Tel. +41 61 683 77 34
Fax +41 61 302 89 18
www.mdpi.com

Mathematics Editorial Office
E-mail: mathematics@mdpi.com
www.mdpi.com/journal/mathematics

www.ingramcontent.com/pod-product-compliance
Lightning Source LLC
LaVergne TN
LVHW070544100526
838202LV00012B/376